云计算与虚拟化技术丛书

The Beginner's Guide to Spring Cloud

极简 Spring Cloud实战

胡劲寒 编著

机械工业出版社
China Machine Press

图书在版编目（CIP）数据

极简 Spring Cloud 实战 / 胡劲寒编著. —北京：机械工业出版社，2019.8
（云计算与虚拟化技术丛书）

ISBN 978-7-111-63281-8

I. 极⋯　II. 胡⋯　III. 互联网络 – 网络服务器　IV. TP368.5

中国版本图书馆 CIP 数据核字（2019）第 148722 号

极简 Spring Cloud 实战

出版发行：机械工业出版社（北京市西城区百万庄大街 22 号　邮政编码：100037）	
责任编辑：高婧雅	责任校对：李秋荣
印　　刷：三河市宏图印务有限公司	版　　次：2019 年 8 月第 1 版第 1 次印刷
开　　本：186mm×240mm　1/16	印　　张：13.75
书　　号：ISBN 978-7-111-63281-8	定　　价：79.00 元

客服电话：（010）88361066　88379833　68326294　　投稿热线：（010）88379604
华章网站：www.hzbook.com　　　　　　　　　　　　读者信箱：hzit@hzbook.com

版权所有 • 侵权必究
封底无防伪标均为盗版
本书法律顾问：北京大成律师事务所　韩光 / 邹晓东

序

在互联网高速发展的时代，谁能够顺应趋势，快速拥抱变化，谁就能在未来的市场充满无限可能性。在滋润的互联网土壤上，各种技术框架、组件得到了蓬勃发展，而微服务无疑是这场技术狂欢中最受关注的热门技术之一。微服务的出现为高速发展的互联网企业带来了新的技术架构理念，松耦合的独立服务组件也使得微服务架构能够快速响应复杂业务的变化，加上对传统软件工程化的革新，也极大推动了自动化发展，以及持续集成与敏捷交付。

但是架构的演进也带来了技术的挑战，特别是服务治理层面的技术复杂性，例如：服务注册与发现、负载均衡、链路跟踪、监控与故障处理（熔断、降级）、APM、请求路由等，一系列的关键技术点都要求技术团队在微服务技术领域中持续投入、持续建设，对于那些缺乏足够技术储备能力的创业团队，技术投入的代价往往过大，如果能有一整套完整的微服务集成与治理的技术方案该有多好。

Spring 作为企业级技术框架中的佼佼者没有错过微服务这个风口。从它第一天出现就注定它的不平凡，Spring 通过其强大的抽象能力以及技术集成能力，结合 Netflix 成熟的开源服务套件，一出现就成为最热门的微服务技术集成方案。

Spring Cloud 也继承了 Spring 一如既往的风格，考虑了微服务的几乎所有功能，另外组件化的思维也为企业微服务架构技术落地提供了更多的灵活性。企业不仅可以通过 Spring Cloud 快速建立起自己的微服务技术体系，也可以通过整合 Spring Cloud 技术组件为已有的技术方案赋能。

劲寒在微服务领域研究多年，有着丰富的微服务落地实施经验，尤其对 Spring Cloud 有着极其深入的研究，在社区上也帮助很多人解决了 Spring Cloud 实际运用中的问题。在互联互通的时代，精华的知识应该开放分享，本着这个理念，劲寒将其对微服务的理解，将 Spring Cloud 研究心得与实战经验全都融合到本书中。

本书不仅介绍了微服务的背景与 Spring Cloud 技术体系价值，让读者快速了解全貌。

更在细节层面对 Spring Cloud 各个组件进行了细致的讲解与深入的原理剖析，最后以微服务技术的视角，结合时下热门的容器化、CI/CD 技术，介绍了微服务未来的运维形态和发展方向，真正做到了让读者既知其然，也知其所以然，更知其未来。

优秀的技术书籍阅读完后让人感觉如同品尝香醇的美酒，口感复杂却层次分明，这本书就带给了我这样的感觉。

——刘洋，同程金服 CTO

前 言

为什么要写这本书

从业近十年，一直从事服务端架构与基础平台等方向的工作，积累了些许服务端架构、微服务领域的心得，此前一直通过技术博客、GitHub、技术社区等方式与同行分享和交流。

Spring Cloud 作为 Spring 推出的一个基于微服务的完整解决方案，能够很方便地使企业进行微服务化转型。笔者结合 Spring Cloud 的微服务落地与推广工作，于实践中更加深了对微服务架构优势的理解。

一次偶然的机会，与策划编辑高婧雅交流后，达成出版意向，开始了写作之旅。

在写作过程中由于工作等个人原因，导致此书的写作计划数次延期，在此郑重对高婧雅编辑表示歉意。幸而在高编辑的不断鞭策与自身的努力下得以成稿。

本书特色

本书力求既全面又精巧，体现在以下方面：揭示 Spring Cloud 的核心特点、关键原理与应用，对于 Spring Cloud 中的每个组件，甚至每个可以支持自定义的扩展场景均有深入介绍。

本书从实战、进阶、全面配置三个层次展开介绍，分为三篇。**基础服务篇**介绍构建一个核心微服务架构不可缺少的部分。**任务与消息篇**则着重介绍 Spring Cloud 针对消息、任务、调用依赖等方面的支持方案。**微服务实战篇**基于 Spring Cloud+Docker 构建一个精简而又五脏俱全的小项目。

读者对象

- 架构师

- 程序开发人员
- 运维管理人员
- 其他对微服务感兴趣的人员

如何阅读本书

本书分为三篇，共 14 章内容。

基础服务篇（第 1～9 章），本篇内容是实践微服务必备的知识点和技能，需要重点学习。

第 1 章对微服务演进历程以及 Spring Cloud 的全貌进行了提纲挈领的介绍，以期读者有全局性认知，使后面的学习不会碎片化。

第 2～8 章主要介绍了在分布式应用中几个核心场景的 Spring Cloud 解决方案，分别深入介绍 Spring Cloud 在服务调用、治理、调用链追踪、熔断及服务网关等方面的实现框架，这些内容是读者实践微服务的基础。

第 9 章主要介绍了 Spring Cloud 中注册中心的其他实现和快速调试、开发脚手架。

任务与消息篇（第 10～13 章），主要介绍消息处理以及任务流依赖处理方面的组件的使用及其实现原理。

微服务实战篇（第 14 章），本篇是基于 Spring Cloud、Docker、OAuth2 构建微服务的一个完整案例。

读者可以根据自身情况，全书阅读或者选择性重点阅读。然而，如果你是一名初学者，请在开始阅读本书之前，先进行一些分布式领域基础理论知识的学习。

勘误和支持

由于笔者水平有限，编写时间仓促，书中难免会出现一些错误或者不准确的地方，恳请读者批评指正。如果你有更多宝贵意见，欢迎发邮件至我的个人邮箱 wawzw123@163.com 进行讨论，我会尽量为读者提供最满意的解答。期待得到你们的真挚反馈，在技术之路上互勉共进。

致谢

感谢 Spring Cloud 官方文档，在写作期间提供给我最全面、最深入、最准确的参考材料，强大的官方文档支持是其他数据库所无法企及的。

感谢 Spring Cloud 中文社区的各位技术专家的博客文章，每次阅读必有所获，本书也多处引用了他们的观点和思想。

感谢所在公司的领导及同事，在微服务技术全面落地的过程中给予的极大信任与支持。

特别致谢

最后，我要特别感谢我的父母和妻子，我为写作这本书牺牲了很多陪伴他们的时间，但也正因为有了他们的付出与支持，我才能坚持写下去。

谨以此书献给我最亲爱的家人，以及众多热爱微服务架构的朋友们！

胡劲寒

目录 Contents

序
前言

第一篇 基础服务篇

第1章 微服务与Spring Cloud 2
1.1 架构演进 2
   1.1.1 服务端架构发展 2
   1.1.2 微服务架构 4
1.2 Spring Cloud 面面观 7
   1.2.1 Spring Cloud 与 Dubbo 对比 7
   1.2.2 Spring Cloud 好在哪里 9
   1.2.3 Spring Cloud 子项目与解决方案 10
1.3 小结 15

第2章 服务发现：Eureka 16
2.1 使用 Eureka 17
   2.1.1 Eureka 服务提供方 18
   2.1.2 Eureka 服务调用方 19
2.2 进阶场景 20
2.3 小结 24

第3章 配置中心：Config 25
3.1 Spring Cloud Config 的组成 25
3.2 使用 Config Server 配置服务端 26
3.3 使用 Config Client 配置客户端 29
3.4 进阶场景 31
   3.4.1 热生效 31
   3.4.2 高可用 32
   3.4.3 安全与加解密 34
   3.4.4 自定义格式文件支持 36
3.5 其他仓库的实现配置 37
3.6 小结 39

第4章 客户端负载均衡：Ribbon 40
4.1 使用 Ribbon 40
4.2 进阶场景 42
   4.2.1 使用配置类 42
   4.2.2 使用配置文件 42
   4.2.3 默认实现 43
4.3 小结 44

第5章 RESTful 客户端：Feign 45
5.1 使用 Feign 45

5.2	进阶场景	46
	5.2.1 配置与默认实现	46
	5.2.2 Feign 整合 Hystrix	47
	5.2.3 数据压缩	48
	5.2.4 日志	48
5.3	小结	49

第 6 章 熔断器：Hystrix 50

6.1	为什么要有熔断	50
6.2	熔断原理	52
6.3	使用 Hystrix	55
6.4	Hystrix 数据监控	58
	6.4.1 健康指示器	58
	6.4.2 监控面板	59
	6.4.3 聚合监控	61
6.5	小结	62

第 7 章 路由网关：Zuul 63

7.1	使用 Zuul	64
7.2	业务场景深入解析	65
7.3	小结	71

第 8 章 网关新选择：Gateway 72

8.1	使用 Gateway	73
8.2	路由断言	76
8.3	过滤器	81
8.4	小结	88

第 9 章 调用链追踪：Spring Cloud Sleuth 89

9.1	术语解释	90
9.2	Zipkin 简介	91

9.3	使用 Zipkin	93
9.4	Span 进阶场景	97
	9.4.1 自定义日志采样策略	97
	9.4.2 Span 的生命周期	98
	9.4.3 重命名 Span	99
	9.4.4 自定义 Span	100
9.5	其他场景与配置	101
9.6	小结	104

第 10 章 加密管理：Vault 105

10.1	初识 HashiCorp Vault	105
10.2	整合 Spring Cloud Vault	111
10.3	认证模式	114
10.4	三方组件支持	116
10.5	小结	118

第 11 章 公共子项目 119

11.1	命令行工具：Spring Boot CLI	119
	11.1.1 安装 Spring Boot CLI	119
	11.1.2 使用 Spring Cloud CLI	120
	11.1.3 加解密	122
11.2	注册中心：Spring Cloud ZooKeeper	122
	11.2.1 安装 ZooKeeper	122
	11.2.2 基于 ZooKeeper 服务发现	122
	11.2.3 相关配置	124
	11.2.4 节点监听	126
11.3	注册中心：Spring Cloud Consul	127
	11.3.1 安装 Consul	127
	11.3.2 基于 Consul 注册服务	127
11.4	小结	128

第二篇 任务与消息篇

第 12 章 消息驱动：Spring Cloud Stream ……………………… 130

- 12.1 Stream 应用模型 …………………… 130
- 12.2 示例 ………………………………… 131
- 12.3 代码解析 …………………………… 133
- 12.4 Spring Integration 支持 …………… 137
- 12.5 Binder 解析 ………………………… 138
- 12.6 常用配置 …………………………… 141
- 12.7 小结 ………………………………… 142

第 13 章 消息总线：Spring Cloud Bus …………………………… 143

- 13.1 使用 Spring Cloud Bus ……………… 144
- 13.2 进阶场景 …………………………… 144
- 13.3 小结 ………………………………… 148

第 14 章 批处理：Spring Cloud Task … 149

- 14.1 使用 Spring Cloud Task …………… 149
- 14.2 进阶场景 …………………………… 150
  - 14.2.1 数据库集成 ………………… 150
  - 14.2.2 任务事件监听 ……………… 152
  - 14.2.3 相关配置项 ………………… 153
  - 12.4.4 整合 Spring Cloud Stream …… 154
- 14.3 源码解析 …………………………… 154
- 14.4 小结 ………………………………… 156

第三篇 微服务实战篇

第 15 章 利用 Docker 进行编排与整合 ………………………… 158

- 15.1 Docker 基础应用 …………………… 158
  - 15.1.1 Docker 基础 ………………… 158
  - 15.1.2 Dockerfile 基础 …………… 159
- 15.2 Spring Cloud 核心组件整合 ……… 161
- 15.3 Dockerfile 编写 …………………… 186
- 15.4 启动与接口测试 …………………… 188
- 15.5 小结 ………………………………… 190

后记 ……………………………………… 191

附录 配置汇总 …………………………… 192

第一篇 Part 1

基础服务篇

本篇将为读者介绍微服务架构的演进过程，带领读者了解什么是微服务，为什么需要微服务，以及微服务与 Spring Cloud 之间是什么关系，为什么要选择 Spring Cloud 来实现微服务而不是市面上现存的其他解决方案。了解之后，相信读者会有自己的答案。

Chapter 1 第 1 章

微服务与 Spring Cloud

本章将带领读者从服务端架构的演进历程开篇，描述服务端架构演进至微服务的必然，同时介绍实现微服务的最佳方式——Spring Cloud，并对 Spring Cloud 的组成及其同业对比的优势进行初步介绍。读者通过本章能够了解微服务是什么，以及 Spring Cloud 如何帮助技术开发人员快速实现微服务。

1.1 架构演进

1.1.1 服务端架构发展

由于人们首先想到的是让两台或多台计算机相互通信，因此构思出了，如图 1-1 所示的简易通信模型。

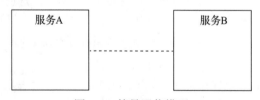

图 1-1　简易通信模型

互相通信的两个服务可以满足最终用户的一些需求。但这个示意图显然过于简单，缺少包括通过代码操作的字节转换和在线路上收发的电信号转换在内的多个层。虽然一定程度上的抽象对于讨论是必需的，但仍需添加网络协议栈（组件）以增加细节内容，如图 1-2 所示。

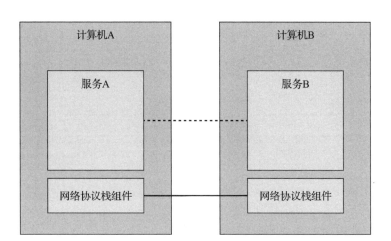

图 1-2　增加网络协议栈（组件）后

上述这个修改过的模型自 20 世纪 50 年代一直使用至今。一开始，计算机很稀少，也很昂贵，所以两个节点之间的每个环节都被精心制作和维护。随着计算机变得越来越便宜，连接的数量和数据量大幅增加。人们越来越依赖网络系统，工程师需要保证他们构建的软件能够达到用户所要求的服务质量。

当然，还有许多问题急需解决以达到用户要求的服务质量。人们需要找到解决方案让机器互相发现，通过同一条线路同时处理多个连接，允许机器在非直连的情况下互相通信，通过网络对数据包进行路由、流量加密等。

其中，有一种机制称为流量控制，下面以此为例。流量控制是一种防止一台服务器发送的数据包超过下游服务器可以承受上限的机制。这是必要的，因为在一个联网的系统中，至少有两台不同的、独立的计算机，彼此之间互不了解。计算机 A 以给定的速率向计算机 B 发送字节，但不能保证 B 可以连续地、以足够快的速度来处理接收到的字节。例如，B 可能正在忙于并行运行其他任务，或者数据包可能无序到达，并且 B 可能被阻塞以等待本应该第一个到达的数据包。这意味着 A 不仅不知道 B 的预期性能，还可能让事情变得更糟，导致 B 过载，B 现在必须对所有这些传入的数据包进行排队处理。

一段时间以来，大家寄希望于建立网络服务和应用程序的开发者能够通过编写代码来解决上面提出的挑战。在这个流程控制示例中，应用程序本身必须包含某种逻辑来确保服务不会因为数据包的原因而过载。这种重联网的逻辑与业务逻辑一样重要。抽象示意图如图 1-3 所示。

随着像 TCP/IP 这样的标准横空出世，流量控制和许多其他问题的解决方案被融入网络协议栈本身。这意味着这些流量控制代码仍然存在，但已经从应用程序转移到了操作系统提供的底层网络层，如图 1-4 所示。

这个模型相当成功。几乎任何一个组织都能够使用商业操作系统附带的 TCP/IP 协议栈来驱动他们的业务，即使有高性能和高可靠性的要求。

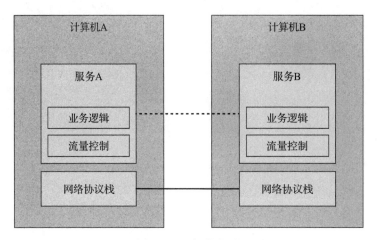

图 1-3　逻辑分离后

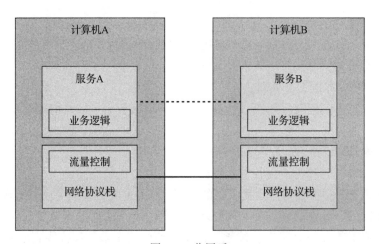

图 1-4　分层后

1.1.2　微服务架构

随着节点和稳定连接的数量越来越多，行业中出现了各种各样的网络系统：从细粒度的分布式代理和对象，到由较大但重分布式组件组成的面向服务的架构。这样的分布式系统带来了几个难题，有一些是新出现的，也有原有难题的"升级版"。

20 世纪 90 年代，Peter Deutsch 和他在 Sun 公司的同事撰写了《分布式计算的八大错误》一文，文中列出了人们在使用分布式系统时通常会做出的一些假设。Peter 认为，这些假设在更原始的网络架构或理论模型中可能是真实存在的，但在现代世界中是不成立的：

- 网络是可靠的；
- 延迟为零；
- 带宽是无限的；

- 网络是安全的;
- 拓扑是不变的;
- 管理员实时监控维护;
- 传输成本为零;
- 网络是同构的。

因此,工程师们必须处理这些问题。

为了处理更复杂的问题,需要转向更加分散的系统(我们通常所说的微服务架构),这在可操作性方面提出了新的要求。下面则列出了必须要处理的问题:

- 计算资源的快速提供;
- 基本的监控;
- 快速部署;
- 易于扩展的存储;
- 可轻松访问边缘;
- 认证与授权;
- 标准化的 RPC;

因此,尽管数十年前开发的 TCP/IP 协议栈和通用网络模型仍然是计算机之间相互通信的有力工具,但更复杂的架构引入了其他层面的问题。此时业界出现了微服务思想,以期解决上述问题。例如,微服务用服务发现与断路器技术来解决上面列出的几个弹性扩展和分布式问题,如图 1-5 所示。

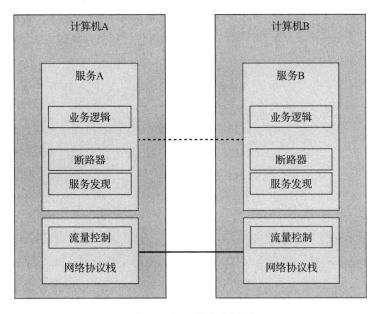

图 1-5 加入微服务层后

微服务架构风格是一种将一个单一应用程序开发为一组小型服务的方法，每个服务运行在自己的进程中，服务间通信采用轻量级通信机制（通常用 HTTP 资源 API）。这些服务围绕业务能力构建并且可通过全自动部署机制独立部署。这些服务共用一个最小型的集中式的管理，服务可用不同的语言开发，使用不同的数据存储技术。

我们为了将系统构建为微服务架构，除了服务是可独立部署、可独立扩展之外，每个服务都提供一个固定的模块边界，甚至允许不同的服务用不同的语言开发，由不同的团队管理。图 1-6 展示了单体应用到微服务的简易图解。

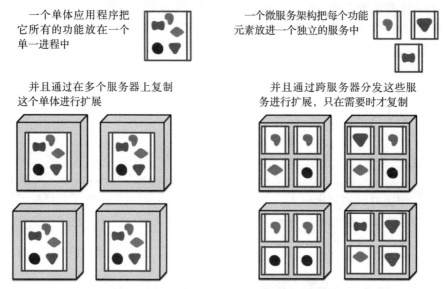

图 1-6　单体应用到微服务的简易图解

然而历史往往会重演，第一批基于微服务构建的系统遵循了与前几代联网计算机类似的策略。这意味着落实上述需求的责任落在了编写服务的工程师身上。我们以服务发现和断路器来说明。

服务发现是在满足给定查询条件的情况下自动查找服务实例的过程，例如，一个名叫 Teams 的服务需要找到一个名为 Players 的服务实例，其中该实例的 environment 属性设置为 production。当用户调用一些提供提供服务发现的组件，它们会返回一个满足条件的服务列表。对于中心化的架构而言，这是一个非常简单的任务，通常可以使用 DNS、负载均衡器和一些端口号的约定（例如，所有服务将 HTTP 服务器绑定到 8080 端口）来实现。而在更分散的环境中，任务开始变得越来越复杂，对于以前可以通过盲目信任 DNS 来查找依赖关系的服务，现在必须处理诸如客户端负载均衡、多种不同环境、地理位置上分散的服务器等问题。如果之前只需要一行代码来解析主机名，那么现在的服务则需要很多行代码来处理由分布式引入的各种问题。

断路器是由 Michael Nygard 在其编写的《Release It》一书中引入的模式，书中对该模式的一些总结：

　　断路器背后的基本思路非常简单。将一个受保护的函数调用包含在用于监视故障的断路器对象中。一旦故障达到一定阈值，则断路器跳闸，并且对断路器的所有后续调用都将返回错误，并完全不接受对受保护函数的调用。通常，如果断路器发生跳闸，还需要对其进行某种监控警报。

　　随着分布式水平的提高，它们也会变得越来越复杂。系统发生错误的概率随着分布式水平的提高呈指数级增长，比如一个组件中的一个故障可能会在许多客户端和客户端的客户端上产生连锁反应，从而触发数千个电路同时跳闸。而且，以前可能只需几行代码就能处理某个问题，现在需要编写大量的代码才能处理。

　　因此微服务反对之声也很强烈，认为微服务增加了系统维护、部署的难度，导致一些功能模块或代码无法复用，增加系统集成与测试的难度，而且随着系统规模增长，会导致系统越来越复杂。那么有没有一种框架或开发平台可以尽可能便捷、一站式解决上述问题呢？有！那便是 Spring Cloud。

1.2　Spring Cloud 面面观

　　为了降低用户构建和维护分布式系统的难度，推动微服务的落地，Spring Cloud 提供了快速构建分布式微服务系统的一些常用功能，如配置管理、服务发现、断路器、智能路由、服务代理、控制总线等提供的一套开发工具。这些工具就相当于分布式系统的样板，Spring Cloud 的使用者可以使用这些样板工具快速构建服务以及相关应用。这些工具能够在任何分布式环境中良好运行，如开发者的计算机、数据中心以及类似 Cloud Foundry 这样的管理平台。

　　Spring Cloud 适用于以下场景：配置管理、服务发现、断路器、智能路由、微代理、控制总线、一次性令牌、全局锁、领导选举、分布式会话、集群状态等。

　　接下来从服务化架构演进的角度来讲述为什么 Spring Cloud 更适应微服务架构。

1.2.1　Spring Cloud 与 Dubbo 对比

　　我们先从 Nginx 说起，了解为什么需要微服务。最初的服务化解决方案是给相同服务提供一个统一的域名，然后服务调用者向这个域发送 HTTP 请求，由 Nginx 负责请求的分发和跳转。

　　这种架构存在很多问题：Nginx 作为中间层，在配置文件中耦合了服务调用的逻辑，这削弱了微服务的完整性，也使得 Nginx 在一定程度上变成了一个重量级的 ESB。图 1-7 标识出了 Nginx 的转发信息流走向。

　　服务的信息分散在各个系统，无法统一管理和维护。每一次的服务调用都是一次尝试，服务消费方并不知道有哪些实例在给他们提供服务。这带来了一些问题：

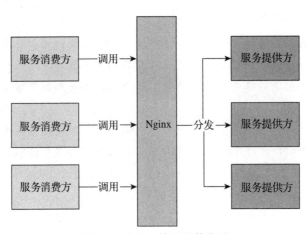

图 1-7 Nginx 转发的信息流

- 无法直观地看到服务提供方和服务消费方当前的运行状况与通信频率;
- 消费方的失败重发、负载均衡等都没有统一策略,这加大了开发每个服务的难度,不利于快速演化。

为了解决上面的问题,我们需要一个现成的中心组件对服务进行整合,将每个服务的信息汇总,包括服务的组件名称、地址、数量等。服务的调用方在请求某项服务时首先通过中心组件获取提供服务的实例信息(IP、端口等),再通过默认或自定义的策略选择该服务的某一提供方直接进行访问,所以考虑引入 Dubbo。

Dubbo 是阿里开源的一个 SOA 服务治理解决方案,文档丰富,在国内的使用度非常高。图 1-8 为 Dubbo 的基本架构图,使用 Dubbo 构建的微服务已经可以较好地解决上面提到的问题。

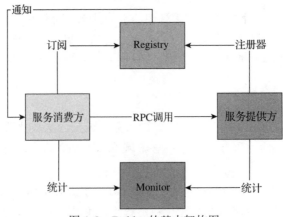

图 1-8 Dubbo 的基本架构图

从图 1-8 中,可以看出:

- 调用中间层变成了可选组件,消费方可以直接访问服务提供方;

- 服务信息被集中到 Registry 中，形成了服务治理的中心组件；
- 通过 Monitor 监控系统，可以直观地展示服务调用的统计信息；
- 服务消费者可以进行负载均衡、服务降级的选择。

但是对于微服务架构而言，Dubbo 并不是十全十美的，也有一些缺陷。

- Registry 严重依赖第三方组件（ZooKeeper 或者 Redis），当这些组件出现问题时，服务调用很快就会中断。
- Dubbo 只支持 RPC 调用。这使得服务提供方与调用方在代码上产生了强依赖，服务提供方需要不断将包含公共代码的 Jar 包打包出来供消费方使用。一旦打包出现问题，就会导致服务调用出错。

笔者认为，Dubbo 和 Spring Cloud 并不是完全的竞争关系，两者所解决的问题域并不一样：Dubbo 的定位始终是一款 RPC 框架，而 Spring Cloud 的目标是微服务架构下的一站式解决方案。如果非要比较的话，Dubbo 可以类比到 Netflix OSS 技术栈，而 Spring Cloud 集成了 Netflix OSS 作为分布式服务治理解决方案，但除此之外 Spring Cloud 还提供了配置、消息、安全、调用链跟踪等分布式问题解决方案。

当前由于 RPC 协议、注册中心元数据不匹配等问题，在面临微服务基础框架选型时 Dubbo 与 Spring Cloud 只能二选一，这也是大家总是拿 Dubbo 和 Spring Cloud 做对比的原因之一。Dubbo 已经适配到 Spring Cloud 生态，比如作为 Spring Cloud 的二进制通信方案来发挥 Dubbo 的性能优势，Dubbo 通过模块化以及对 HTTP 的支持适配到 Spring Cloud。

1.2.2 Spring Cloud 好在哪里

作为新一代的服务框架，Spring Cloud 提出的口号是开发"面向云的应用程序"，它为微服务架构提供了更加全面的技术支持。结合我们一开始提到的微服务的诉求，参见表 1-1，把 Spring Cloud 与 Dubbo 进行一番对比。

表 1-1 Spring Cloud 与 Dubbo 功能对比

功能名称	Dubbo	Spring Cloud
服务注册中心	ZooKeeper	Spring Cloud Netflix Eureka
服务调用方式	RPC	REST API
服务网关	无	Spring Cloud Netflix Zuul
断路器	不完善	Spring Cloud Netflix Hystrix
分布式配置	无	Spring Cloud Config
服务跟踪	无	Spring Cloud Sleuth
消息总线	无	Spring Cloud Bus
数据流	无	Spring Cloud Stream
批量任务	无	Spring Cloud Task

Spring Cloud 抛弃了 Dubbo 的 RPC 通信，采用的是基于 HTTP 的 REST 方式。严格来

说,这两种方式各有优劣。虽然从一定程度上来说,后者牺牲了服务调用的性能,但也避免了上面提到的原生 RPC 带来的问题。而且 REST 相比 RPC 更为灵活,服务提供方和调用方,不存在代码级别的强依赖,这在强调快速演化的微服务环境下显得更加合适。

很明显,Spring Cloud 的功能比 Dubbo 更加强大,涵盖面更广,而且作为 Spring 的拳头项目,它也能够与 Spring Framework、Spring Boot、Spring Data、Spring Batch 等其他 Spring 项目完美融合,这些对于微服务而言是至关重要的。前面提到,微服务背后一个重要的理念就是持续集成、快速交付,而在服务内部使用一个统一的技术框架,显然比将分散的技术组合到一起更有效率。更重要的是,相比于 Dubbo,它是一个正在持续维护的、社区更加火热的开源项目,这就可以保证使用它构建的系统持续地得到开源力量的支持。

下面列举 Spring Cloud 的几个优势。

- Spring Cloud 来源于 Spring,质量、稳定性、持续性都可以得到保证。
- Spirng Cloud 天然支持 Spring Boot,更加便于业务落地。
- Spring Cloud 发展得非常快,从开始接触时的相关组件版本为 1.x,到现在将要发布 2.x 系列。
- Spring Cloud 是 Java 领域最适合做微服务的框架。
- 相比于其他框架,Spring Cloud 对微服务周边环境的支持力度最大。对于中小企业来讲,使用门槛较低。

1.2.3　Spring Cloud 子项目与解决方案

我们从整体上来看一下 Spring Cloud 各个组件如何配套使用,如图 1-9 所示。

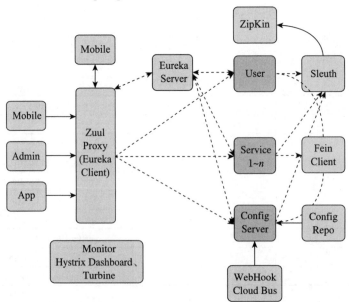

图 1-9　Spring Cloud 各个组件协同示意图

从图 1-9 可以看出，Spring Cloud 各个组件相互配合，合作支持了一套完整的微服务架构。

Spring Cloud 从设计之初就考虑了绝大多数互联网公司架构演化所需的功能，这些功能都是以插拔的形式提供，方便合理选择需要的组件进行集成，从而使用过程中更加平滑、顺利。

Spring Cloud 提供了标准化的、一站式的技术方案。下面按照 Spring Cloud 子项目和其提供的分布式方案来分别了解 Spring Cloud 价值。

系统一旦走向分布式，其复杂程度成倍增长，传统单体应用那种只考虑业务逻辑的开发方式已经不再适用。正因其复杂性，目前只有业务需求大的大型互联网公司才会（被迫）采用，而且需要投入大量的技术力量来开发基础设施，也造成了小公司"用不起"分布式架构的情况。现在这一局面正在逐渐被打破，因为 Netflix 开源了经过实战考验的一系列基础设施构件，加上 Spring Cloud 的大力支持，开发分布式系统已经不再像以前那样可怕了。

（1）服务的注册发现和软负载均衡

Spring Cloud Netflix 通过 Eureka Server 实现服务注册中心，通过 Ribbon 实现软负载均衡。

（2）分区治理

Eureka 支持 Region 和 Zone 的概念。其中一个 Region 可以包含多个 Zone。Eureka 在启动时需要指定一个 Zone 名，即当前 Eureka 属于哪个 Zone，如果不指定则属于 defaultZone。Eureka Client 也需要指定 Zone，Client（当与 Ribbon 配置使用时）在向 Server 获取注册列表时会优先向自己所在 Zone 的 Eureka 发请求，只有自己 Zone 中的 Eureka 全挂了才会尝试向其他 Zone 发请求。当获取到远程服务列表后，Client 也会优先向同一个 Zone 的服务发起远程调用。Region 和 Zone 可以对应于现实中的大区和机房，如在华北地区有 10 个机房，在华南地区有 20 个机房，那么分别为 Eureka 指定合理的 Region 和 Zone 能有效避免跨机房调用，而一个地区的 Eureka 坏掉也不会导致整个该地区的服务都不可用。

（3）Ribbon 软负载均衡

Ribbon 在工作时分成两步：第一步先选择 Eureka Server，优先选择同一个 Zone 且负载较少的 Server，第二步再根据用户指定的策略，在从 Server 取到的服务注册列表中选择一个地址。其中 Ribbon 默认提供了三种策略：轮询、断路器和根据响应时间加权。

（4）声明式 HTTP RESTful 客户端 Feign

Feign 与 Apache Http Client 这类客户端最大的不同，是它允许通过定义接口的形式构造 HTTP 请求，不需要手动拼参数，使用起来与正常的本地调用没有什么区别：

```
@FeignClient(name = "ea")
public interface AdvertGroupRemoteService {
    @RequestMapping(value = "/group/{groupId}", method = RequestMethod.GET)
    AdvertGroupVO findByGroupId(@PathVariable("groupId") Integer adGroupId);
}
```

这里只需要调用 AdvertGroupRemoteService.findByGroupId() 方法就能完成向目标主机发送 HTTP 请求并封装返回结果的效果。

（5）断路器

断路器（Cricuit Breaker）是一种能够在远程服务不可用时自动熔断（打开开关），并在远程服务恢复时自动恢复（闭合开关）的设施，Spring Cloud 通过 Netflix 的 Hystrix 组件提供断路器、资源隔离与自我修复功能。

（6）资源隔离

Hystrix 对每一个依赖服务都配置了一个线程池，对依赖服务的调用会在线程池中执行。例如，我们设计服务 I 的线程池大小为 20，那么 Hystrix 会最多允许有 20 个容器线程调用服务 I，如果超出 20，Hystrix 会拒绝并快速失败。这样即使服务 I 长时间未响应，容器最多也只能堵塞 20 个线程，剩余 80 个线程仍然可以处理用户请求。

（7）快速失败

快速失败是防止资源耗尽的关键一点。当 Hystrix 发现在过去某段时间内对服务 I 的调用出错率达到某个阈值时，Hystrix 就会"熔断"该服务，后续任何发向服务 I 的请求都会快速失败，而不是白白让调用线程去等待。

（8）自我修复

处于熔断状态的服务，在经过一段时间后，Hystrix 会让其进入"半关闭"状态，即允许少量请求通过，然后统计调用的成功率。如果这个请求能成功，Hystrix 会恢复该服务，从而达到自我修复的效果。其中，在服务被熔断到进入半关闭状态之间的时间，就是留给开发人员排查错误并恢复故障的时间，开发人员可以通过监控措施得到提醒并线上排查。

（9）监控方案

监控是保障分布式系统健康运行必不可少的方案。基于 Spring Cloud，我们可以从两个纬度进行监控：Hystrix 断路器的监控和每个服务监控状况的监控。

图 1-10 是 Hystrix 提供的 Dashboard 图形化监控。

可见图 1-10 中监控信息应有尽有（调用成功率、平时响应时间、调用频次、断路器状态等）。可以通过编程的方式定时获取该信息，并在断路器熔断时通过短信、邮件等方式通知开发者。

Hystrix 的监控数据默认保存在每个实例的内存中，Spring Boot 提供了多种方式，可以导入到 Redis、TSDB 以供日后分析使用。

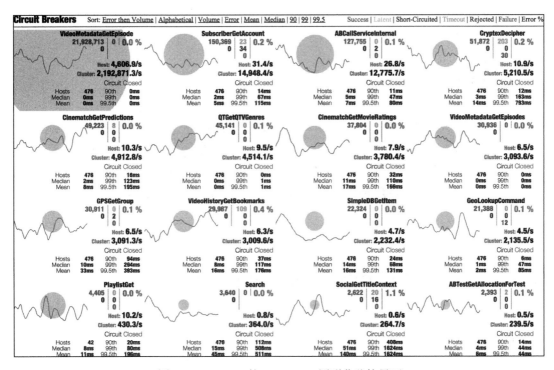

图 1-10　Hystrix 的 Dashboard 图形化监控界面

除此之外，Spring Cloud 还提供了对单个实例的监控，如图 1-11 所示。其中包含了接口调用频次、响应时间、JVM 状态、动态日志等各种开发者关心的信息。

Spring Cloud 的版本介绍

（1）版本命名

之前提到过，Spring Cloud 是一个拥有诸多子项目的大型综合项目，原则上其子项目也都维护着自己的发布版本号。每一个 Spring Cloud 的版本都会包含不同的子项目版本，为了管理每个版本的子项目清单，避免版本名与子项目的发布号混淆，没有采用版本号的方式，而是通过命名的方式。这些版本采用了伦敦地铁站的名字，根据字母表的顺序来对应版本时间顺序，比如：最早的 Release 版本为 Angel，第二个 Release 版本为 Brixton，依此类推。

（2）版本号

经过上面的解释，不难猜出，Angel.SR6 和 Brixton.SR5 中的 SR6、SR5 就是版本号了。

当一个版本的 Spring Cloud 项目的发布内容积累到临界点或者一个严重 bug 解决后，就会发布一个"service releases"版本，简称 SRX 版本，其中 X 是一个递增数字。

（3）当前版本

我们在选择 Spring Boot 与 Spring Cloud 版本的时候，还是需要尽可能按照 Spring

Cloud 官方版本依赖关系来选择：
- Brixton 版本对应 Spring Boot 1.3.x；
- Camden 版本对应 Spring Boot 1.4.x；
- Dalston 版本对应 Spring Boot 1.5.x。

图 1-11　监控界面

在开始正式介绍每个项目之前，需要准备如表 1-2 所示的环境。

表 1-2 必备工具

工　具	版本或描述
JDK	1.7 及以上版本
IDE	IntelliJ IDEA、Eclipse 等
Maven	3.x
Spring Boot	1.5 及以上版本

1.3 小结

微服务架构将一个应用拆分成多个独立的、具有业务属性的服务，每个服务运行在不同的进程中，服务与服务之间通过轻量级的通信机制互相协作、互相配合，从而为终端用户提供业务价值。因此，微服务架构强调的是一种独立开发、独立测试、独立部署、独立运行的高度自治的架构模式，也是一种更灵活、更开放、更松散的演进式架构。

希望通过本章所介绍的微服务的定义、核心特征、优缺点，以及微服务实现框架 Spring Cloud 的综合介绍，大家能够对微服务以及 Spring Cloud 有全面的了解，接下来的章节将逐一、详细介绍 Spring Cloud 的各个组件。

第 2 章

服务发现：Eureka

服务发现是微服务架构中的一个重要概念。试想当系统服务之间的依赖越来越多，A 服务可能需要调用 B、C、D 等服务，同时被调用方也就是服务提供方可能为了保证自身高可用，还需要同时以集群的模式部署 B1、B2、C1、C2 等，想象一下 A 的配置文件该有多复杂。将服务提供方的地址写死在配置文件中，那么服务提供方如果横向扩展增加实例，是不是还需要修改作为服务调用方的 A 的配置文件？这时我们就迫切需要一种服务发现的机制。所有的服务提供方启动时向注册中心报告自身的信息，包括自己的地址、端口，以及提供哪些服务等相关信息。当服务调用方需要调用服务时，只需要问注册中心是谁提供了相关的服务，注册中心返回哪些提供方提供了这些服务，调用方就可以自己根据注册中心返回的信息去请求了。Eureka 就提供了这样一种能力，同时自身作为注册中心的同时也提供了高可用的支持，支持集群部署时各个节点之间的注册数据同步复制。

Eureka 是 Netflix 开源的一款提供服务注册和发现的产品，提供了完整的服务注册和服务发现实现，也是 Spring Cloud 体系中最重要、最核心的组件之一。

通俗讲，Eureka 就是一个服务中心，将所有可以提供的服务都注册到它这里来管理，其他各调用者需要的时候去注册中心获取，然后服务调用方再向服务提供方发起调用，避免了服务之间的直接调用，方便后续的水平扩展、故障转移等。

所以，服务中心这么重要的组件一旦宕机将会影响全部服务，因此需要搭建 Eureka 集群来保持高可用性，建议生产中最少配备两台。随着系统流量的不断增加，需要根据情况来扩展某个服务，Eureka 内部提供均衡负载的功能，只需要增加相应的服务端实例即可。那么在系统的运行期间某个实例宕机怎么办？Eureka 提供心跳检测机制，如果某个实例在规定的时间内没有进行通信则会被自动剔除掉，避免了某个实例挂掉而影响服务。

因此，使用 Eureka 就自动具有了注册中心、负载均衡、故障转移的功能。

它主要包括两个组件。
- Eureka Client：一个 Java 客户端，用于简化与 Eureka Server 的交互（通常就是微服务中的客户端和服务端）。
- Eureka Server：提供服务注册和发现的能力（通常就是微服务中的注册中心）。

各个微服务启动时，会通过 Eureka Client 向 Eureka Server 注册自己，Eureka Server 会存储该服务的信息，如图 2-1 所示。

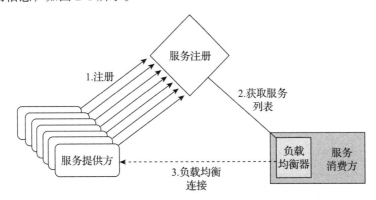

图 2-1　服务注册与获取的交互流程

也就是说，每个微服务的客户端和服务端都会注册到 Eureka Server，这就衍生出了微服务相互识别的话题。
- 同步：每个 Eureka Server 同时是 Eureka Client（逻辑上的），多个 Eureka Server 之间通过复制的方式完成服务注册表的同步，从而实现 Eureka 的高可用。
- 识别：Eureka Client 会在本地缓存 Eureka Server 中的信息。

即使所有 Eureka Server 节点都宕掉，服务消费方仍可使用本地缓存中的信息找到服务提供方。
- 续约：微服务会周期性（默认 30s）地向 Eureka Server 发送心跳以续约（Renew）自己的信息（类似于 heartbeat 机制）。
- 续期：Eureka Server 会定期（默认 60s）执行一次失效服务检测功能，它会检查超过一定时间（默认 90s）没有续约的微服务，发现则会注销该微服务节点。

当一个注册器客户端通过 Eureka 进行注册时，它会带上一些描述自己情况的元数据，如地址、端口、健康指示器地址、主页等。Eureka 会接收每一个服务实例发送的心跳包。如果心跳包超过配置的间隔时间，则这个服务实例就会被移除。

2.1　使用 Eureka

接下来尝试一个 Eureka 的示例。

新建一个 Maven 项目，在 pom.xml 中添加对 Eureka 服务端的依赖：

```xml
<dependency>
    <groupId>org.springframework.cloud</groupId>
    <artifactId>spring-cloud-starter-eureka-server</artifactId>
</dependency>
```

新建 EurekaServerApplication.java，添加 @EnableEurekaServer 注解：

```java
@SpringBootApplication
@EnableEurekaServer
public class EurekaServerApplication {
    public static void main(String[] args) {
        SpringApplication.run(EurekaServerApplication.class, args);
    }
}
```

启动程序后，打开 http://localhost:8080/ 就可以看到如图 2-2 所示的监控页面，图中展示了向 Eureka 注册的所有客户端。

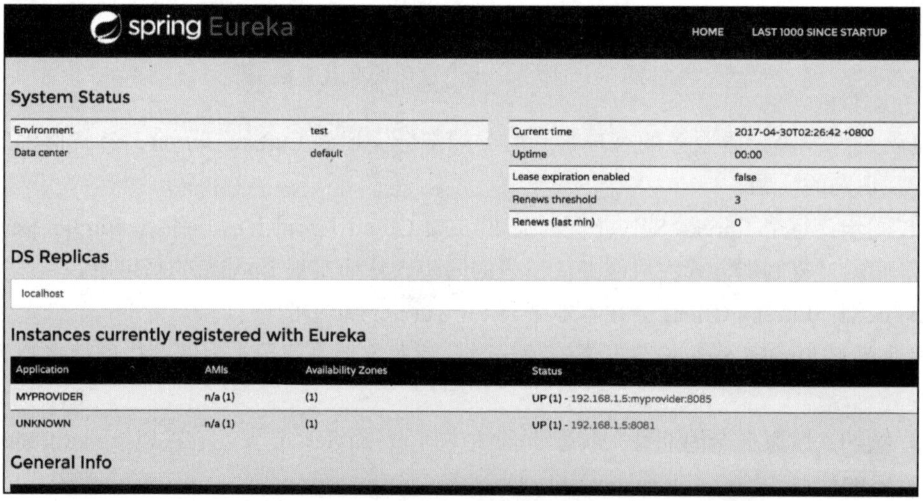

图 2-2　服务注册 Eureka 的监控显示结果

2.1.1　Eureka 服务提供方

启动一个服务提供方并注册到 Eureka。新建一个项目并在 pom.xml 中添加依赖：

```xml
<dependency>
    <groupId>org.springframework.cloud</groupId>
    <artifactId>spring-cloud-starter-eureka</artifactId>
</dependency>
<dependency>
    <groupId>org.springframework.boot</groupId>
```

```xml
<artifactId>spring-boot-starter-web</artifactId>
</dependency>
```

新建 EurekaProviderApplication.java，并对外暴露一个 sayHello 的 HTTP 接口。

```java
@SpringBootApplication
@EnableEurekaClient
@RestController
public class EurekaProviderApplication {
    @RequestMapping("/sayHello")
    public  String sayHello(String name){
        return "hello "+name;
    }
    public static void main(String[] args) {
        new SpringApplicationBuilder(EurekaProviderApplication.class).web(true).
            run(args);
    }
}
```

在 application.yml 配置 Eureka 服务端地址以及自身服务名称：

```yaml
eureka:
    client:
        serviceUrl:
            defaultZone: http://localhost:8761/eureka/ #在注册中心配置Eureka地址
spring:
    application:
        name: myprovider #自身名字
```

启动 EurekaProviderApplication 后，再去看 Eureka 的监控页面，应该能看到如图 2-3 所示信息，表明服务已经注册到 Eureka。

Application	AMIs	Availability Zones	Status
MYPROVIDER	n/a (1)	(1)	UP (1) - 192.168.1.5:myprovider:8085

图 2-3　在 Eureka 上注册成功

 注意 还有一种方法就是直接使用原生的 com.netflix.discovery.EurekaClient（对应 Spring Cloud 的 DiscoveryClient）。通常，尽量不要直接使用原始的 Netflix 的 Eureka Client，因为 Spring 已经对其进行封装抽象，应尽可能使用 DiscoveryClient。

2.1.2　Eureka 服务调用方

一旦在应用中使用了 @EnableDiscoveryClient 或者 @EnableEurekaClient，就可以从 Eureka Server 中使用服务发现功能。

利用如下代码，通过 DiscoveryClient 可以从 Eureka 服务端获得所有提供 myprovider 服务的实例列表，并根据获得的 ServiceInstance 对象获取每个提供方的相关信息，如端口

IP 等，从中选取第一个提供方，通过 Spring 的 RestTemplate 发起 HTTP 请求进行调用。对于获取到的提供方集合，我们根据什么规则去选取哪个提供方？能否做到负载均衡呢？这将在第 5 章详细介绍。

```java
@SpringBootApplication
@EnableDiscoveryClient
@Slf4j
@RestController
public class EurekaClientApplication {
    @Bean
    public RestTemplate restTemplate(){
        return new RestTemplate();
    }
    @Autowired
    private DiscoveryClient client;
    @Autowired
    private RestTemplate restTemplate;
    @RequestMapping(value = "/sayHello" ,method = RequestMethod.GET)
    public String sayHello(String name) {
        List<ServiceInstance> instances = client.getInstances("myprovider");
        if (!instances.isEmpty()) {
            ServiceInstance instance = instances.get(0);
            log.info(instance.getUri().toString());
            String result=restTemplate.getForObject(instance.getUri().toString()+
                "/sayHello?name="+name,String.class);
            return  result;
        }
        return "failed";
    }
    public static void main(String[] args) {
        SpringApplication.run(EurekaClientApplication.class, args);
    }
}
```

此时，我们通过 HTTP 请求调用 /sayHello 接口，则可以看到服务调用方从 Eureka 获取服务提供方信息，并进行调用的日志信息了。

 提示　不要在 @PostConstruct 方法以及 @Scheduled 中（或者任何 ApplicationContext 还没初始完成的地方）使用 EurekaClient。其需要等待 SmartLifecycle（phase=0）初始化完成才可以。

2.2　进阶场景

尝试过 Eureka 的基本使用场景后，虽然功能能够实现，但是落实到具体的企业级应用场景时，必然会有许多需要自定义的配置以及一些生产环境可能需要用到的参数。下面列出一些场景供读者参考。

（1）Eureka 的健康检查

默认情况下，Eureka 通过客户端心跳包来检测客户端状态，并不是通过 spring-boot-actuator 模块的 /health 端点来实现的。默认的心跳实现方式可以有效地检查 Eureka 客户端进程是否正常运作，但是无法保证客户端应用能够正常提供服务。由于大多数微服务应用都会有一些外部资源依赖，比如数据库、Redis 缓存等，如果应用与这些外部资源无法连通时，实际上已经不能提供正常的对外服务了，但因为客户端心跳依然在运行，所以它还是会被服务消费者调用，而这样的调用实际上并不能获得预期的效果。当然，我们可以开启 Eureka 的健康检查，这样应用状态就可以同步给 Eureka 了。在 application.yml 中添加如下配置即可：

```
eureka:
  client:
    healthcheck:
      enabled:true
```

Eureka 中的实例一共有如下几种状态：UP、DOWN、STARTING、OUT_OF_SERVICE、UNKNOWN。如果需要更多的健康检查控制，可以实现 com.netflix.appinfo.HealthCheckHandler 接口，根据自己的场景进行操作。

 eureka.client.healthcheck.enabled=true 只能在 application.yml 中设置，如果在 bootstrap.yml 中设置，会导致 Eureka 注册为 UNKNOWN 的状态。

（2）自我保护模式

Eureka 在设计时，认为分布式环境的网络是不可靠的，可能会因网络问题导致 Eureka Server 没有收到实例的心跳，但是这并不能说明实例宕了，所以 Eureka Server 默认会打开保护模式，它主要是网络分区场景下的一种保护。

一旦进入保护模式，Eureka Server 将会尝试保护其服务注册表中的信息，不再删除里面的数据（即不会注销任何微服务）。在这种机制下，它仍然鼓励客户端再去尝试调用这个所谓 down 状态的实例，当网络故障恢复后，该 Eureka Server 节点会自动退出自我保护模式。若确实调用失败，熔断器就派上用场了。

关于熔断器，第 6 章会详细介绍并演示。

通过修改注册中心的配置文件 application.yml，即可打开或关闭注册中心的自我保护模式：

```
eureka:
  server:
    enable-self-presaervation: false        #关闭自我保护模式（默认为打开）
```

综上，自我保护模式是一种应对网络异常的安全保护措施。它的理念是宁可同时保留所有实例（健康实例和不健康实例都会保留），也不盲目注销任何健康的实例。使用自我保护模式，可以让 Eureka 集群更加健壮、稳定。

（3）踢出宕机节点

自我保护模式打开时，已关停节点是一直显示在 Eureka 首页的。

关闭自我保护模式后，由于其默认的心跳周期比较长等原因，要过一会儿才会发现已关停节点被自动踢出了。若想尽快踢出，就只能修改默认的心跳周期参数了。

注册中心的配置文件 application.yml 需要修改的地方如下所示：

```
eureka:
    server:
        enable-self-preservation: false         # 关闭自我保护模式（默认为打开）
        eviction-interval-timer-in-ms: 1000     # 续期时间，即扫描失效服务的间隔时间
                                                （默认为60*1000ms）
```

客户端的配置文件 application.yml 需要修改的地方为：

```
eureka:
    instance:
        lease-renewal-interval-in-seconds: 5        # 心跳时间，即服务续约间隔时间（默认为30s）
        lease-expiration-duration-in-seconds: 15    # 发呆时间，即服务续约到期时间（默认为90s）
    client:
        healthcheck:
            enabled: true                           #开启健康检查（依赖spring-boot-starter-actuator）
```

 更改 Eureka Server 的更新频率将打破注册中心的自我保护功能，不建议生产环境自定义这些配置。

（4）注册服务慢的问题

客户端去注册中心默认持续 30s，直到实例自身、服务端、客户端各自元数据本地缓存同步完成后服务才可用（至少需要 3 次心跳周期）。可以通过 eureka.instance.leaseRenewalIntervalInSeconds 修改这个周期，改善客户端链接到服务的速度。不过，考虑短期的网络波动以及服务续期等情况，在生产环境最好用默认设定。

（5）服务状态显示 UNKNOWN

如果在 Eureka 监控页面发现服务状态显示 UNKNOWN，则很大可能是把微服务的 eureka.client.healthcheck.enabled 属性配置在 bootstrap.yml 里面的问题。比如，实际测试发现，Eureka 首页显示的服务状态，本应是 UP(1)，却变成大红色的粗体 UNKNOWN(1)。

（6）自定义 InstanceId

两个相同的服务（端口不同），如果注册时设置的都是 ${spring.application.name}，那么 Eureka 首页只会看到一个服务名字，而无法区分有几个实例注册上来了。于是，可以自定义生成 InstanceId 的规则。

Eureka 服务名默认如下：

```
${spring.cloud.client.hostname}:${spring.application.name}:${spring.application.
    instance_id:${server.port}}
```

可以在配置文件中通过 eureka.intance.intance_id 来自定义：

```
eureka:
    instance:
        #修改显示的微服务名为：IP:端口
        instance-id: ${spring.cloud.client.ipAddress}:${server.port}
```

（7）自定义 Eureka 控制台服务的链接

既然微服务显示的名称允许修改，那么其对应的点击链接也是可以修改的。

同样，还是修改微服务的配置文件，如下所示：

```
eureka:
    instance:
        # ip-address: 192.168.6.66      #只有prefer-ip-address=true时才会生效
        prefer-ip-address: true         #设置微服务调用地址为IP优先（默认为false）
```

Eureka 首页显示的微服务调用地址，默认是 http://hostName:port/info。

而在设置 prefer-ip-address=true 之后，调用地址会变成 http://ip:port/info。

这时若再设置 ip-address=192.168.6.66，则调用地址会变成 http://192.168.6.66:2100/info。

（8）健康度指示器

一个 Eurake 实例的状态页和健康指示器默认为 /info 和 /health，这两个是由 Spring Boot Actuator 应用提供的访问端点。可以通过以下方式进行修改：

```
application.yml
eureka:
    instance:
        statusPageUrlPath: ${management.context-path}/info
        healthCheckUrlPath: ${management.context-path}/health
```

这些地址会被用于 Eureka 对客户端元数据的获取，以及健康检测。

（9）就近原则

用户量比较大或者用户地理位置分布范围很广的项目，一般都会有多个机房。这个时候如果上线服务的话，我们希望一个机房内的服务优先调用同一个机房内的服务，当同一个机房的服务不可用时，再去调用其他机房的服务，以达到减少延时的作用。Eureka 有 Region 和 Zone 的概念，可以理解为现实中的大区（Region）和机房（Zone）。Eureka Client 在启动时需要指定 Zone，它会优先请求自己 Zone 的 Eureka Server 获取注册列表。同样，Eureka Server 在启动时也需要指定 Zone。如果没有指定，其会默认使用 defaultZone。

（10）高可用配置

Eureka Server 也支持运行多实例，并以互相注册的方式（即伙伴机制）来实现高可用的部署，即每一台 Eureka 都在配置中指定另一个 Eureka 地址作为伙伴，它在启动时会向伙伴节点获取注册列表。如此一来，Eureka 集群新加机器时，就不用担心注册列表的完整性。所以，只需要在 Eureka Server 里面配置其他可用的 serviceUrl，就实现了注册中心的高可用。

我们新建两个 Eureka 服务端项。第 1 个服务端项为 EurekaServer1 项目中的配置文件 /src/main/resources/application.yml。

```
server:
    port: 8989
eureka
    serviceUrl:
        defaultZone:
```

第 2 个服务端项为 EurekaServer2 项目中的配置文件 /src/main/resources/application.yml。

```
server:
    port: 9898
eureka
    serviceUrl:
        defaultZone:
```

启动 Server1、Server2 后，分别访问 http://127.0.0.1:8989/eureka/、http://127.0.0.1:9898/eureka/，发现 DS Replicas、General Info 模块出现了对方的信息。读者可以自行测试，分别单独向 Server1 或者 Server2 进行服务注册时，都会自动同步给另外一个注册中心。

在生产环境中大于两台注册中心的条件下，可以同理将其配置成如图 2-4 所示的双向环形。

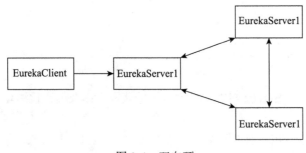

图 2-4　双向环

2.3　小结

本章详细讲解了 Eureka 组件的工作原理，并结合示例介绍了 Eureka 的服务提供方和服务调用方的使用步骤。在进阶场景章节中我们深入探究了各个场景下使用 Eureka 的各种解决方案与自定义场景的配置。下个章开始学习如何使用 Config 组件来对 Eureka 以及其他组件进行动态化配置。

第 3 章 Chapter 3

配置中心：Config

对于传统的单体应用，配置文件可以解决配置问题，但是当多机部署时，修改配置依然是烦琐的问题。

在微服务中，由于系统拆分的粒度更小，微服务的数量比单体应用要多得多（基本上多一个数量级），通过配置文件来管理配置变得更不可行。

所以，对于微服务架构而言，一个通用的分布式配置管理是必不可少的。在大多数微服务系统中，都会有一个名为"配置文件"的功能模块来提供统一的分布式配置管理。

在研发流程中有测试环境、UAT 环境、生产环境等隔离，因此每个微服务又对应至少三个不同环境的配置文件。这么多的配置文件，如果需要修改某个公共服务的配置信息，如缓存、数据库等，难免会产生混乱，这时就需要引入 Spring Cloud 的另外一个组件：Spring Cloud Config。

Spring Could Config 是一个提供了分布式配置管理功能的 Spring Cloud 子项目。在以往的单体应用中往往是代码与配置文件放在一个应用包中，但是随着系统的体量越来越大，我们会将系统分成多个服务，对于这么多服务的配置管理以及热生效等方面的支持将会越来越麻烦。Spring Cloud Config 完美解决了这些问题。

在市面上有一些开源产品，如百度的 DisConf、淘宝的 Diamond，以及很多基于 ZooKeeper 的各个公司自主开发的产品。这些产品可能由于某些问题已经停止维护，导致文档资料不全、重复造轮子等各种问题。而 Spring Cloud Config 由于可与 Spring 无缝集成、功能强大、社区活跃等各方面原因，成为开发中不可不着重考虑的一项技术。

3.1 Spring Cloud Config 的组成

Spring Cloud Config 项目提供了如下的功能支持：

- 提供服务端和客户端支持；
- 集中式管理分布式环境下的应用配置；
- 基于 Spring 环境，与 Spring 应用无缝集成；
- 可用于任何语言开发的程序；
- 默认实现基于 Git 仓库，可以进行版本管理；
- 可替换自定义实现；
- Spring Cloud Config Server 作为配置中心服务端；
- 拉取配置时更新 Git 仓库副本，保证是最新结果；
- 支持数据结构丰富，包括 yml、json、properties 等；
- 配合 Eureka 可实现服务发现，配合 Spring Cloud Bus 可实现配置推送更新；
- 配置存储基于 Git 仓库，可进行版本管理；
- 简单可靠，有丰富的配套方案；
- Spring Cloud Config Client 提供（如 SVN、Local 等）开箱即用的客户端实现；
- Spring Boot 项目不需要改动任何代码，加入一个启动配置文件指明使用 Config Server 中哪个配置文件即可。

下面分别从配置仓库、Config Server、Config Client 的使用与概念解释，以及 Config Server 的高可用、全局通知、安全性、加解密等方面来介绍 Spring Cloud Config。

3.2 使用 Config Server 配置服务端

本节先使用 Git 作为配置文件存储仓库，后文中会介绍使用 SVN、本地目录以及自行扩展等方式。

首先，我们需要在以 Maven 作为依赖管理的项目 pom.xml 中添加 spring-cloud-starter-config、spring-cloud-config-server 两项依赖，以及以 spring-boot-starter-parent 作为父项目。

```xml
<dependencyManagement>
    <dependencies>
        <dependency>
            <groupId>org.springframework.cloud</groupId>
            <artifactId>spring-cloud-dependencies</artifactId>
            <version>Edgware.RELEASE</version>
            <type>pom</type>
            <scope>import</scope>
        </dependency>
    </dependencies>
</dependencyManagement>
<dependencies>
    <dependency>
        <groupId>org.springframework.cloud</groupId>
        <artifactId>spring-cloud-starter-config</artifactId>
```

```xml
        </dependency>
        <dependency>
            <groupId>org.springframework.cloud</groupId>
            <artifactId> spring-cloud-config-server</artifactId>
        </dependency>
</dependencies>
```

在项目中创建 ConfigServerApplication 类，其中 @EnableConfigServer 注解表示允许该服务以 HTTP 形式对外提供配置管理服务。

```
@SpringBootApplication
@EnableConfigServer
public class ConfigServerApplication {
    public static void main(String[] args) {
        SpringApplication.run(ConfigServerApplication.class, args);
    }
}
```

添加 application.yml，新增如下内容指定 Git 仓库的地址：

```
server:
    port: 8888
spring:
    cloud:
config:
application:
    name: myConfigServer
        server:
            git:
                #Git仓库地址
            uri: https://git.oschina.net/wawzw123/SpringCloudBookCode.git
            search-paths: config-repo
```

如下为配置文件中的配置项。

1）spring.cloud.config.server.git.uri：配置 Git 仓库位置。

2）spring.cloud.config.server.git.searchPaths：配置仓库路径下的相对搜索位置，可以配置多个。

3）spring.cloud.config.server.git.username：访问 Git 仓库的用户名。

4）spring.cloud.config.server.git.password：访问 Git 仓库的用户密码。

读者在自行测试的时候需要自行创建 Git 仓库，并根据 Git 仓库信息自行替换 application.properties 中的内容。我们已经事先在实例的仓库中添加了如下几个文件，用于进行不同分支的不同 key 的读取测试。

在 Master 分支中添加如下文件和内容。

1）configServerDemo.properties：key1=master-default-value1；

2）configServerDemo-dev.properties：key1=master-dev-value1；

3）configServerDemo-test.properties：key1=master-test-value1；

4）configServerDemo-prd.properties：key1=master-prd-value1。

在 Branch 分支中添加如下文件和内容。

1）configServerDemo.properties：key1=branch-prd-value1；

2）configServerDemo-dev.properties：key1=branch-dev-value1；

3）configServerDemo-test.properties：key1=branch-test-value1；

4）configServerDemo-prd.properties：key1=branch-prd-value1。

在服务端启动后，可以基于 HTTP 请求访问如下 URL 进行配置获取。可以通过如下几种格式的 HTTP 向配置中心发起配置文件获取的请求：

1）/{application}/{profile}[/{label}]；

2）/{application}-{profile}.yml；

3）/{application}-{profile}.json；

4）/{label}/{application}-{profile}.yml；

5）/{label}/{application}-{profile}.json；

6）/{application}-{profile}.properties；

7）/{label}/{application}-{profile}.properties。

- application：应用名称，默认取 spring.application.name，也可以通过 spring.cloud.config.name 指定。
- profile：环境，如 dev（开发）、test（测试）、prd（生产）等，通过 spring.cloud.config.profile 指定。
- label：版本，对应不同的分支，通过 spring.cloud.config.label 指定。

比如，尝试通过 curl 或者浏览器等发起 HTTP 请求"http://localhost:8888/configServerDemo/test/master"，将会得到如下响应内容。

```
{
    "name": "configServerDemo",
    "profiles": [
        "test"
    ],
    "label": null,
    "version": "32d326655ae7d17be752685f29d017ba42e8541a",
    "propertySources": [
        {
            "name": "https://git.oschina.net/wawzw123/Spring CloudBookCode.git/
                config-repo/configServerDemo-test.properties",
            "source": {
                "key1": "master-test-value1"
            }
        },
        {
"name":"https://git.oschina.net/wawzw123/Spring CloudBookCode.git/config-repo/
    configServerDemo.properties",
```

```
            "source": {
                "key1": "master-default-value1"
            }
        }
    ]
}
```

访问 http://localhost:8888/configServerDemo-test.yml，则会得到如下结果：

```
key1: master-test-value1
```

在尝试了手动从配置中心获取配置项之后，我们接下来尝试启动一个示例项目来自动从配置中心获取配置项。

3.3 使用 Config Client 配置客户端

接下来创建一个 Spring Boot 应用作为配置管理的客户端来读取 Config Server 中提供的配置，如图 3-1 所示。

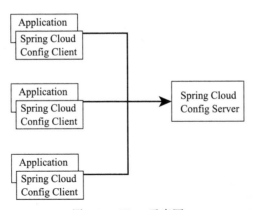

图 3-1　Client 示意图

在 pom.xml 中添加如下依赖：

```xml
<dependency>
    <groupId>org.springframework.cloud</groupId>
    <artifactId>spring-cloud-starter-config</artifactId>
</dependency>
<dependency>
    <groupId>org.springframework.boot</groupId>
    <artifactId>spring-boot-starter-test</artifactId>
    <scope>test</scope>
</dependency>
```

添加 bootstrap.yml：

```yaml
server:
    port: 7002
spring:
    application:
        name: cloudConfigDemo${server.port}
    cloud:
        config:
            profile: dev
            label: master
            name: configServerDemo
            uri: http://localhost:8888/
```

> **注意** 上面这些属性必须配置在 bootstrap.properties 中，config 部分才能被正确加载。因为 config 的相关配置会先于 application.properties，而 bootstrap.properties 的加载也先于 application.properties。

创建 ConfigClientApplication 类并添加 @EnableAutoConfiguration 注解，表示自动获取项目中的配置变量：

```java
@SpringBootApplication
@RestController
@EnableAutoConfiguration
public class ConfigClientApplication {
@Value("${key1}")
String foo;
@RequestMapping("/say")
@ResponseBody
public String say(){
    return foo;
}
    public static void main(String[] args) {
        SpringApplication.run(Application.class, args);
    }
}
```

在 ConfigClientApplication 中我们创建了一个 RestController 提供 Web 服务，输出读取到的 key1 的值。

启动 main 方法，看到控制台信息中将有如下一行日志，包含了配置的相关信息：

```
[main]c.c.c.ConfigServicePropertySourceLocator:Located-environment:name=conf
    igServerDemo,profiles=[test],label=branch1,version=575f8f8ded872700c7abc
    fb6bbbecf02f9271a17, state=null
```

现在我们尝试访问 http://localhost:8084/say，会得到 branch1-test-value1 的响应。

下面我们来一起了解 Spring Cloud Config Client 可能用到的常用配置。

（1）客户端快速失败

有的时候，需要在 Config Server 连接不上时直接启动失败。需要这个特性时可以设置

bootstrap 配置项 spring.cloud.config.failFast=true 来开启。

（2）客户端重试

可以在 Config Server 不可用时，让客户端重试。可以通过设置"spring.cloud.config.failFast=false;"在 classpath 中增加 spring-retry、spring-boot-starter-aop 依赖。默认情况下会重试 6 次，每次间隔 1000ms 并以 1.1 乘以次数方式递增。也可以通过 'spring.cloud.config.retry' 系列配置来修改相关配置。而且我们可以自己实现一个 RetryOperationsInterceptor 来详细地自定义重试策略。

（3）HTTP 权限

如果要对 HTTP 请求进行账号密码的权限控制，可以配置服务器 URI 或单独的用户名和密码属性，bootstrap.yml 配置文件如下：

```
spring:
  cloud:
    config:
      uri: https://user:secret@myconfig.mycompany.com
```

或者：

```
spring:
  cloud:
    config:
      uri: https://myconfig.mycompany.com
      username: user
      password: secret
```

> 注意：spring.cloud.config.password 和 spring.cloud.config.username 值覆盖 URI 中提供的内容。

3.4 进阶场景

3.4.1 热生效

在应用运行时经常会有修改配置的需求，那么在 Spring Cloud Config 中如何让修改 Git 仓库的配置动态生效呢？我们在 ConfigClientApplication 类上加上 @RefreshScope 注解并在 Config Client 的 pom.xml 中添加 spring-boot-starter-actuator 监控模块，其中包含了 /refresh 刷新 API，并启动 Config Client。如下为 pom 文件中的依赖项：

```xml
<dependency>
    <groupId>org.springframework.boot</groupId>
    <artifactId>spring-boot-starter-actuator</artifactId>
</dependency>
```

在配置完之后，我们进行如下尝试来验证配置时进行了热生效。

步骤1 访问 http://localhost:8084/say 得到响应"key1: master-test-value1"。

步骤2 将 configServerDemo-test.properties 的内容修改为 master-test-value2 并通过 Git 提交对配置文件的修改。

步骤3 请求 Config Client 的 http://localhost:8084/say，依旧是"key1: master-test-value1"。因为 Client 未接到任何通知进行本地配置更新。

步骤4 通过 POST 访问 http://localhost:8084/refresh 得到响应 ["key1"]，表明 key1 已被更新。

步骤5 访问 http://localhost:8084/say，相应内容已经变成了 master-test-value2。

然而，如果每次修改了配置文件就要手动请求 /refresh，这一定不是我们想要的效果。在第 11 章中我们将介绍如何使用 Bus 来通知 Spring Cloud Config，如图 3-2 所示。

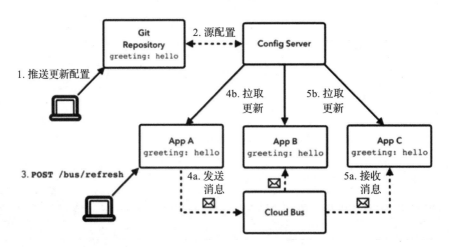

图 3-2 Spring Cloud Bus 与 Spring Cloud Config 交互流程

3.4.2　高可用

在上文中我们以在配置文件中指定配置中心 Config Server 的实例地址的方式来定位 Config Server。一旦遇到 Config Server 宕机，Config Client 将无法继续获取配置，且对 Config Server 进行横向扩展时也需要修改每一个 Config Client 的配置文件。当单台 Config Server 压力过大时，客户端也无法做到负载均衡。

Spring Cloud 针对 Config Server 同样支持通过 Eureka 进行服务注册的方式。我们将 Config Server 的所有实例以服务提供方的形式注册在 Eureka 上。Config Client 以服务消费方的形式去 Eureka 获取 Config Server 的实例，这样也就同时支持由 Eureka 组件提供的故障转移、服务发现、负载均衡等功能。

我们接下来对之前的 Config Server 进行改造，在 Config Server 的 Maven pom.xml 上增

加 Eureka 的依赖：

```xml
<dependency>
    <groupId>org.springframework.cloud</groupId>
    <artifactId>spring-cloud-starter-eureka</artifactId>
</dependency>
```

在配置文件 application.yml 中追加 Eureka 的注册中心地址的配置：

```yaml
eureka:
    client:
        serviceUrl:
            defaultZone: http://localhost:8989/eureka/
```

在 ConfigServerApplication.java 主类中标注 @EnableEurekaClient 注解。

接下来，启动 Config Server，并查看 Eureka 控制台，可以看见已经注册的服务提供方。之后需要让 Config Client 去 Eureka 获取 Config Server 地址。我们来对 Config Server 进行改造。

在 Maven 中新增对 Eureka 的依赖：

```xml
<dependency>
    <groupId>org.springframework.cloud</groupId>
    <artifactId>spring-cloud-starter-eureka</artifactId>
</dependency>
```

在 bootstrap.yml 中新增 Eureka 注册中心地址的配置，并去掉 spring.cloud.config.uri 的静态指定。通过 spring.cloud.config.discovery.enabled 指定打开 Spring Cloud Config 的动态发现服务功能的开关，通过 spring.cloud.config.discovery.serviceId 指定在注册中心的配置中心中的 ServiceId。

```yaml
eureka:
    client:
        serviceUrl:
            defaultZone: http://localhost:8989/eureka/
spring:
    application:
        name: cloudConfigDemo${server.port}
    cloud:
        config:
            profile: dev
            label: master
            name: configServerDemo
            discovery:
                enabled: true
            service-id: myConfigServer
            #uri: http://localhost:8888/
```

在 ConfigClientApplication.java 主类中增加 @EnableDiscoveryClient 注解，使其打开服务发现客户端功能。

启动 Config Client 时可以发现，如果配置服务部署多份，通过负载均衡，可以实现高可用。

3.4.3 安全与加解密

使用 Spring Cloud Config 时，可能有些场景需要将配置服务暴露在公网或者其他需要加权限安全控制的场景。可以使用 Spring Security 来整合 Spring Cloud Config。

使用 Spring Boot 默认的基于 HTTP 安全方式，仅仅需要引入 Spring Security 依赖（如：可以通过 spring-boot-starter-security）。引入此依赖的默认情况是使用一个用户名和一个随机产生的密码，这种方式并不是很靠谱，因此，建议通过 spring-boot-starter-security 配置密码，并对其进行加密处理。

在默认情况下，启动 Config Server 时，会看到启动日志中有如下类似信息：

```
b.a.s.AuthenticationManagerConfiguration :
Using default security password: 7bbc28c2-b60f-4996-8eb5-87b4f57e976c
```

这就是 Spring Security 默认生成的密码，同样也可以通过配置文件自定义账号和密码：

```
security:
  user:
    name: testuser
    password: testpassword
```

1. 服务端加解密

在实际生产环境使用过程中，就算加入了账号和密码等方式的权限控制，数据存储在 Config Server 中依旧可能被泄露，那么可以对数据加密后再存储在 Config Server 中。

如果远程资源是一个经过加密的内容（以 {cipher} 开头），在发送给客户端之前运行时会被解密。这样，配置内容就不用明文存放了。

我们先来使用 JDK 自带的 keytool 工具生成加解密时所需要用到的密钥：

```
keytool
-genkey
-alias cloudtest （别名）
-keypass 123456 （别名密码）
-keyalg RSA （算法）
-validity 365 （有效期，天）
-keystore mykey.keystore （指定生成证书的位置和证书名称）
-storepass mypass （获取keystore信息的密码）
```

接下来需要填入一些无关紧要的信息：

```
你的名字与姓氏是什么？
    [Unknown]:  hjh
你的组织单位名称是什么？
    [Unknown]:  spring
```

```
你的组织名称是什么?
  [Unknown]:  spring
你所在的城市或区域名称是什么?
  [Unknown]:  shanghai
你所在的省/市/自治区名称是什么?
  [Unknown]:  cn
该单位的双字母国家/地区代码是什么?
  [Unknown]:  cn
CN=hjh, OU=spring, O=spring, L=shanghai, ST=cn, C=cn是否正确?
  [否]:  y
```

接下来将密钥信息复制进 Resources 目录并配置进 Config Server 配置文件中:

```
encrypt:
    key-store:
        location: classpath:mykey.keystore
        password: mypass
        alias: cloudtest
        secret: 123456application.yml
```

在 location 中也可以使用 file:// 来配置文件路径。

接下来尝试访问 http://localhost:8888/encrypt 并以 POST 方式提交需要加密的内容。

```
$ curl localhost:8888/encrypt -d mysecret
AQATZVzrgr9M+doCEiRdL44JD2rB+A2HzX/I6Sec6w04+VW+znApTHZoiJhL0Fn4+3u73aUi
    5euejvokwmAx+ttBPX8UrhxMcDHZmqj1ADm2XAqX1/NEJtkcfVSFCrkyAztzlT/u+6/
    uzHRUMZhiJDn41yYtGKtt9/zlni9WKcEBxhSb2XMYuJL21EL2q4w2rD9awLYfJBy4MD6fbPC2mlZ
    0XCFuCDR7mslneLQtB/bkKcVUR/p5g8GJ8qWUt9T6DGQ52QgxTCoRvJcUFzulRD+A3b4UhuHmumd
    P0i7wM+hnTI+6h/HXVZ33Ju8SGRtnYXp7Bnz69T4NPZRT7Ov6S/4/IJMObwrSNSfZv7tAV2BSRj4
    U6xhBCCAcXdVrTHQzlpM=
```

请求返回的内容就是服务端根据我们配置的密钥加密后的结果。

同样,以 POST 方式请求 http://localhost:8888/decrypt 并传入密文,将会返回解密后的结果。

```
$ curl localhost:8888/decrypt -d AQATZVzrgr9M+doCEiRdL44JD2rB+A2HzX/I6Sec6w0
    4+VW+znApTHZoiJhL0Fn4+3u73aUi5euejvokwmAx+ttBPX8UrhxMcDHZmqj1ADm2XAqX1/
    NEJtkcfVSFCrkyAztzlT/u+6/uzHRUMZhiJDn41yYtGKtt9/zlni9WKcEBxhSb2XMYuJL21EL2q
    4w2rD9awLYfJBy4MD6fbPC2mlZ0XCFuCDR7mslneLQtB/bkKcVUR/p5g8GJ8qWUt9T6DGQ52Qgx
    TCoRvJcUFzulRD+A3b4UhuHmumdP0i7wM+hnTI+6h/HXVZ33Ju8SGRtnYXp7Bnz69T4NPZRT7Ov
    6S/4/IJMObwrSNSfZv7tAV2BSRj4U6xhBCCAcXdVrTHQzlpM=
asdasd
```

> **注意** 如果在请求 /encrypt 和 /decrypt 的时候服务端抛出"Illegal key size"异常,则表明 JDK 中没有安装 Java Cryptography Extension。
> Java Cryptography Extension(JCE)是一组包,提供用于加密、密钥生成和协商以及 MAC(Message Authentication Code)算法的框架和实现,提供对对称、不对称、块和流密码的加密支持,还支持安全流和密封的对象。它不提供对外出口,用它开

发并完成封装后将无法调用。

下载地址为 http://www.oracle.com/technetwork/java/javase/downloads/index.html，下载并解压完成后，将其复制到 JDK/jre/lib/security 中即可。

在实际使用过程中，只需要将生成好的密文以 {cipher} 开头填入配置中即可。

```
spring:
  datasource:
    username: dbuser
    password: '{cipher}FKSAJDFGYOS8F7GLHAKERGFHLSAJ'
```

如果使用 properties 格式配置文件，则加密数据不要加上双引号。可以在 application.properties 中加入如下配置：

```
spring.datasource.username: dbuser
spring.datasource.password: {cipher}FKSAJDFGYOS8F7GLHAKERGFHLSAJ
```

这样就可以安全共享此文件，同时可以保护密码。

2. 客户端解密

有的时候需要客户端自行对密文进行解密，而不是在服务端解密。这就需要明确指定配置数据在服务端发出时不解密：spring.cloud.config.server.encrypt.enabled=false。

3.4.4 自定义格式文件支持

在某些场景可能不总是以 YAML、Properties、JSON 等格式获取配置文件，可能内容是自定义格式的，希望 Config Server 将其以纯文本方式来处理而不做其他加工。Config Server 提供了一个访问端点 /{name}/{profile}/{label}/{path} 来支持这种需求，这里的 {path} 是指文件名。

当资源文件被找到后，与常规的配置文件一样，也会先处理占位符。例如，在上文案例的 Git 仓库中上传：

```
nginx.conf
server {
    listen 80;
    mykey ${key1};
}
```

接下来重启 Config Server 并请求 Nginx 的配置文件。如尝试请求 http://localhost:8888/configServerDemo/dev/master/nginx.conf，将会得到如下响应：

```
server {
    listen 80;
    mykey master-dev-value-dev;
}
```

可以看到正常匹配了我们上传的自定义格式文件，并替换了占位符。

3.5　其他仓库的实现配置

1. 配置 Git

在应用配置文件与特定配置文件中可以通过正则表达式来支持更为复杂的情况。在 {application}/{profile} 中可以使用通配符进行匹配，如果有多个值可以使用逗号分隔，配置文件示例如下：

```
spring:
  cloud:
    config:
      server:
        git:
          uri: https://github.com/spring-cloud-samples/config-repo
          repos:
            simple: https://github.com/simple/config-repo
            special:
              pattern: special*/dev*,*special*/dev*
              uri: https://github.com/special/config-repo
            local:
              pattern: local*
              uri: file:/home/configsvc/config-repo
```

如果 {application}/{profile} 没有匹配到任何资源，则使用 spring.cloud.config.server.git.uri 配置的默认 URI。

上面例子中 pattern 属性是一个 YAML 数组，也可以使用 YAML 数组格式来定义。这样可以设置成多个配置文件，示例如下：

```
spring:
  cloud:
    config:
      server:
        git:
          uri: https://github.com/spring-cloud-samples/config-repo
          repos:
            development:
              pattern:
                - */development
                - */staging
              uri: https://github.com/development/config-repo
            staging:
              pattern:
                - */qa
                - */production
              uri: https://github.com/staging/config-repo
```

每个资源库有一个可选的配置，用来指定扫描路径，示例如下：

```
spring:
```

```yaml
cloud:
  config:
    server:
      git:
        uri: https://github.com/spring-cloud-samples/config-repo
        searchPaths: foo,bar*
```

这样系统就会自动搜索 foo 的子目录,以及以 bar 开头的文件夹中的子目录。

默认情况下,当第一次请求配置时,系统复制远程资源库。系统也可以配置成一启动就复制远程资源库,示例如下:

```yaml
spring:
  cloud:
    config:
      server:
        git:
          uri: https://git/common/config-repo.git
          repos:
            team-a:
              pattern: team-a-*
              cloneOnStart: true
              uri: http://git/team-a/config-repo.git
            team-b:
              pattern: team-b-*
              cloneOnStart: false
              uri: http://git/team-b/config-repo.git
            team-c:
              pattern: team-c-*
              uri: http://git/team-a/config-repo.git
```

上面的例子中 team-a 的资源库会在启动时就从远程资源库进行复制,其他的则等到第一次请求时才从远程资源库复制。

2. 配置权限与 HTTPS

如果远程资源库设置了权限认证,则可以如下配置:

```yaml
spring:
  cloud:
    config:
      server:
        git:
          uri: https://github.com/spring-cloud-samples/config-repo
          username: trolley
          password: strongpassword
```

如果不使用 HTTPS 和用户认证,可以使用 SSH URI 的格式。例如,git@github.com:configuration/cloud-configuration,这就需要先有 SSH 的 key。这种方式系统会使用 JGit 库进行访问,可以去查看相关文档。可以在 ~/.git/config 中设置 HTTPS 代理配置,也可以通过 JVM 参数 -Dhttps.proxyHost、-Dhttps.proxyPort 来配置代理。

提示 用户不知道自己的 ~/.git 目录时，可以使用 git config --global 来指定。例如：git config --global http.sslVerify false。

3. 配置 SVN

如果希望使用 SVN 充当配置仓库来替换 Git，配置也与 Git 类似，同样支持账户、密码、搜索路径等配置，这里不再赘述，SVN 配置示例如下：

```
spring:
  cloud:
    config:
      server:
        svn:
          uri: https://subversion.assembla.com/svn/spring-cloud-config-repo/
        #git:
        # uri: https://github.com/pcf-guides/configuration-server-config-repo
        default-label: trunk
profiles:
  active: subversion
```

4. 配置本地仓库

如果希望配置仓库从本地 classpath 或者文件系统加载配置文件，可以通过 spring.profiles.active=native 开启。默认从 classpath 中加载，如果使用 "file:" 前缀加载文件系统，则从本地路径中加载。当然也可以使用 ${} 样式的环境占位符，例如：file:///${user.home}/config-repo。

3.6 小结

有了 Spring Cloud Config，可以实现对任意一个集成过的 Spring 程序进行动态化参数配置及热生效等相关操作，从而实现程序与配置隔离，解耦编码与环境之间的耦合。这同样是微服务架构所需要的。接下来我们将进入下一章开始学习服务端之间的调用。

第 4 章
客户端负载均衡：Ribbon

Ribbon 是一个基于 HTTP 和 TCP 客户端的负载均衡器。它可以在客户端配置 ribbonServerList（服务端列表），然后轮询请求以实现负载均衡。它在联合 Eureka 使用时，ribbonServerList 会被 DiscoveryEnabledNIWSServerList 重写，扩展成从 Eureka 注册中心获取服务端列表。同时它会用 NIWSDiscoveryPing 来取代 IPing，它将职责委托给 Eureka 来确定服务端是否已经启动。

4.1 使用 Ribbon

这个实例会涉及一个 Eureka 注册中心服务端、两个服务提供方、一个服务消费方。注册中心将继续使用上文中已经实现过的 Eureka。启动两个 myprovider，包含 SayHello 功能的服务提供方并注册在 Eureka 上。

接下来，对上文中 Eureka 客户端的代码进行改造。

```
@SpringBootApplication
@EnableDiscoveryClient
@EnableEurekaClient
@Slf4j
@RestController
public class EurekaClientApplication2 {
    @Bean
    public RestTemplate restTemplate() {
        return new RestTemplate();
    }
    @Autowired
    private LoadBalancerClient client;
```

```
    @Autowired
    private RestTemplate restTemplate;
    @RequestMapping(value = "/sayHello" ,method = RequestMethod.GET)
    public String sayHello(String name) {
        ServiceInstance instance =  client.choose("myprovider");
        log.info(instance.getUri().toString());
        String result=restTemplate.getForObject(instance.getUri().toString()+"/
            sayHello?name="+name,String.class);
        return  result+"from "+instance.getPort();
    }
    public static void main(String[] args) {
        SpringApplication.run(EurekaClientApplication2.class, args);
    }
}
```

将 DiscoveryClient 改为 LoadBalancerClient，并调用其 choose 方法，会使原先的得到一个 ServiceInstance 集合变为得到单个 ServiceInstance 实例。

之后，返回 ServiceInstance 的端口来区分我们的两个服务提供方，接下来用浏览器请求 http://localhost:8081/sayHello?name=test。

```
hello testfrom 8084
hello testfrom 8085
```

得到的 HTTP 响应为交替出现 8085 和 8084。这表明 Ribbon 客户端正在为轮询方式实现负载均衡。

还有一种更简便的方法，标注 @RibbonClient 并设置需要获取的服务名，并给 RestTemplate 标注上 @LoadBalance 注解。可以直接通过 RestTemplate 使用 URLhttp://myprovider/ 进行请求，Ribbon 同样有助于实现负载均衡。

```
@SpringBootApplication
@EnableDiscoveryClient
@EnableEurekaClient
@Slf4j
@RestController
@RibbonClient(name = "myprovider")
public class EurekaClientApplication2 {
    @Bean
    @LoadBalanced
    public RestTemplate restTemplate() {
        return new RestTemplate();
    }
    @Autowired
    private RestTemplate restTemplate;
    @RequestMapping(value = "/sayHello" ,method = RequestMethod.GET)
    public String sayHello(String name) {
        String result=restTemplate.getForObject("http://myprovider/sayHello?name=
            "+name,String.class);
        return  result;
```

```
    }
    public static void main(String[] args) {
        SpringApplication.run(EurekaClientApplication2.class, args);
    }
}
```

4.2 进阶场景

上文展示的一个实例实现了 Robbin 的基本功能，但是默认的一些参数和策略可能无法满足用户的所有场景，如用户可能想自定义 HTTP 线程池大小，或者想将负载均衡策略从轮询换成随机。这就需要自定义配置。

4.2.1 使用配置类

在使用 @RibbonClient 时，可以设置 configuration 的值来自定义配置类，实现对 Ribbon 的策略等进行配置，还可以通过自定义配置 MyRibbonConfiguration 使 PingUrl 替换默认的 NoOpPing。

```
@RibbonClient(name = "myprovider",configuration = MyRibbonConfiguration.class)
@Configuration
public class MyRibbonConfiguration {
    @Bean
    public IPing ribbonPing(IClientConfig config) {
        return new PingUrl();
    }
}
```

> **注意** 需要确保 MyRibbonConfiguration 不会被 @Configuration 或 @ComponentScan 扫描到，否则就会被所有 @RibbonClients 共享（自动注入）。如果使用 @ComponentScan 或 @SpringBootApplication 时，也需要避免 MyRibbonConfiguration 被自动扫描。

4.2.2 使用配置文件

用户也可以通过在配置文件中用 <client>.ribbon.* 来配置 Ribbon 客户端相关属性，如 myprovider. ribbon. PoolMaxThreads=50，配置属性可以参见 CommonClientConfigKey 类。

若需要在 users 服务中指定一个 IRule，可以在配置文件 application.yml 中配置如下内容：

```
users:
    ribbon:
        NFLoadBalancerRuleClassName: com.netflix.loadbalancer.WeightedResponseTimeRule
```

还有一些可能需要用到的配置项如下。

- NFLoadBalancerClassName：指定一个 ILoadBalancer 实现类；
- NFLoadBalancerRuleClassName：指定一个 IRule 实现类；
- NFLoadBalancerPingClassName：指定一个 IPing 实现类；
- NIWSServerListClassName：指定一个 ServerList 实现类；
- NIWSServerListFilterClassName：指定一个 ServerListFilter 实现类。

 这些属性指定的类，会覆盖使用 @RibbonClient(configuration=MyRibbonConfig.class) 注解中的定义，同时也会覆盖 Spring Cloud Netflix 的默认策略。

4.2.3 默认实现

查看 @RibbonClient 的源码注释，可知 org.springframework.cloud.netflix.ribbon.RibbonClientConfiguration 为其默认的配置类。读者可以自行查看想要了解的默认实现类。

列举可能需要更加关注的配置项及其默认实现，参见表 4-1。

表 4-1　Ribbon 中的默认实现

Bean 类型	Bean 名称	类　　名	备　　注
IClientConfig	ribbonClientConfig	DefaultClientConfigImpl	Ribbon 客户端设置的默认值
IRule	ribbonRule	ZoneAvoidanceRule	负载均衡策略，以哪种方式从提供者列表中选取
IPing	ribbonPing	NoOpPing	检查服务端是否通畅的方式
ServerList<Server>	ribbonServerList	ConfigurationBasedServerList	服务列表可以选择静态配置或者动态获取等方式
ServerListFilter<Server>	ribbonServerListFilter	ZonePreferenceServerListFilter	服务列表过滤器
ILoadBalancer	ribbonLoadBalancer	ZoneAwareLoadBalancer	负载均衡器

下面列出 SpringCloud 提供的一些针对 IRule 可以开箱即用的默认实现。

- **RoundRobinRule**：轮询，每一次来自网络的请求轮流分配给内部服务器，从 1 至 N，然后重新开始。此种均衡算法适合服务器组中的所有服务器都有相同的软硬件配置并且平均服务请求相对均衡的情况。
- **AvailabilityFilteringRule**：一种根据可用性进行过滤的策略。默认情况下，如果 RestClient 最后三次连接失败，则实例将电路跳闸。一旦一个实例电路跳闸，它将保持在这种状态 30s，之后电路再次被启用。然而，如果连接失败，它将再次变为"电路跳闸"，并且其等待开启的时间将以连续故障的数量呈指数增长。
- **WeightedResponseTimeRule**：对于该规则，每个服务器根据其平均响应时间给予权重。响应时间越长，权重就越少。

下面是 SpringCloud 提供的一些针对 ServerList 的开箱即用的默认实现。

- StaticServerList：只能支持 BaseLoadBalancer 或其子类通过 BaseLoadBalancer.setServersList() 设置的服务器列表。
- ConfigurationBasedServerList：默认实现，可以通过配置文件 <clientName>.ribbon.listOfServers 来设置。如果属性动态更改，服务器列表也将针对负载平衡器进行更改，如 sample-client.ribbon.listOfServers=www.microsoft.com:80,www.yahoo.com:80, www.google.com:80。
- DiscoveryEnabledNIWSServerList：该 ServerList 实现从 Eureka 客户端获取服务器列表。对于 ServerListFilter 是为 DynamicServerListLoadBalancer 过滤从 ServerList 实现中获取的服务列表的过滤器。
- ZoneAffinityServerListFilter：过滤掉与客户端不是同一个 Zone 的服务器，除非客户端 Zone 中没有可用的服务列表。
- ServerListSubsetFilter：此过滤器确保客户端仅看到由 ServerList 实现返回的整个服务列表的固定子集。它还可以定期用新服务列表替代可用性差的子集中的服务器。

4.3 小结

通过本章的学习，可以了解到 Ribbon 作为负载均衡器对高性能、高可用等方面起到的不可忽视的作用。当然作为一个独立的组件，同样可以将它集成进需要负载均衡的任何场景，这需要对 Ribbon 有深入理解，相信读者通过本章的学习，这应该不是一件困难的事情。

第 5 章 RESTful 客户端：Feign

在实际开发过程中，尽管 Eureka 的注册发现和 Ribbon 的客户端负载均衡很强大，但是我们不可能每次都对 URL 进行远程调用，像拼参数这种事情会让每个开发人员痛苦不堪，而 Feign 将会解决这些问题。Feign 是一个 Web 服务的客户端框架，它让 Web 服务的客户端开发变得更容易。只需要使用 Feign 创建一个接口加上一个注解就行了。在使用 Feign 时，Spring Cloud 还可以整合 Ribbon 和 Eureka，为 HTTP 客户端提供负载均衡的能力。

5.1 使用 Feign

上文在 Eureka 服务调用方示例的 pom 文件中加入 Feign 的 Maven 依赖：

```
<dependency>
    <groupId>org.springframework.cloud</groupId>
    <artifactId>spring-cloud-starter-feign</artifactId>
</dependency>
```

定义 Feign 接口类：

```
@FeignClient(name = "MYPROVIDER")
public interface IBizApi {
    @RequestMapping("/sayHello")
    public String sayHello(@RequestParam("name") String name);
}
```

然后在主类上标注 @EnableFeignClients。

```
@SpringBootApplication
```

```java
@EnableFeignClients
@EnableEurekaClient
@Slf4j
@RestController
public class FeignClientApplication {
    @Autowired
    private IBizApi bizApi;
    @RequestMapping("/sayHello")
    public String sayHello(){
        return  bizApi.sayHello("haha");
    }
    public static void main(String[] args) {
        SpringApplication.run(FeignClientApplication.class, args);
    }
}
```

这样就能很方便地把一个 HTTP 的请求方式转换为编码友好的 Java 接口形式。

5.2 进阶场景

5.2.1 配置与默认实现

就像 Ribbon 一样，也可以通过 @FeignClient 中的 configuration 来指定用户的自定义配置类。

```
@FeignClient(name = "MYPROVIDER",configuration = MyFeignConfiguration.class)
```

在 MyFeignConfiguration 类中，定义了 Gson 来实现序列化与反序列化。

```java
public class MyFeignConfiguration {
    @Bean
    public Decoder decoder() {
        return new GsonDecoder();
    }
    @Bean
    public Encoder encoder() {
        return new GsonEncoder();
    }
}
```

当然，需要先将 Gson 的类库加入 Maven 依赖。

```xml
<dependency>
    <groupId>io.github.openfeign</groupId>
    <artifactId>feign-gson</artifactId>
    <version>9.4.0</version>
</dependency>
```

如果想针对所有 FeignClient 进行配置，可以通过 @EnableFeignClients 的 defaultConfiguration

属性来制定。这些配置会被应用到所有的 Feign 客户端上。

org.springframework.cloud.netflix.feign.FeignClientsConfiguration 是 Feign 的默认配置类，可以看到 Feign 对一些接口默认指定了哪些实现，参见表 5-1。

表 5-1　配置类解释及其默认实现

类　　型	Bean 名称	类　　名	备　　注
Decoder	feignDecoder	ResponseEntityDecoder	HTTP 响应报文转换为对象时的反序列化器
Encoder	feignEncoder	SpringEncoder	与上相反
Logger	feignLogger	Slf4jLogger	日志实现
Contract	feignContract	SpringMvcContract	用于验证哪些接口和值被允许定义在接口上
Feign.Builder	feignBuilder	HystrixFeign.Builder	客户端构造器
Client	feignClient	LoadBalancerFeignClient	指定具体发起 HTTP 请求的客户端。如果启用 Ribbon，则是 LoadBalancer-FeignClient，否则使用 Feign 作为默认客户端

下面列出 SpringCloud 提供的一些针对 Encoder、Decoder、Contract、Client 开箱即用的默认实现。

（1）Encoder、Decoder
- Gson：一个以 JSON 格式进行转换的类库，提供了 GsonEncoder 和 GsonDecoder。
- Jackson：一个以 JSON 格式进行转换的类库，提供了 JacksonEncoder 和 JacksonDecoder。
- Sax：XML 格式的解码器，提供了 SaxDecoder。
- JAXB：XML 格式的转换类库，提供了 JAXBEncoder 和 JAXBDecoder。

（2）Contract
- JAXRSContract：使得 Feign 支持 JAX-RS 规范定义的注解，如 @GET @Path 等。
- SpringMvcContract：使得 Feign 支持 Spring MVC 注解，如 @RequestMapping 等。
- HystrixDelegatingContract：支持 Hystrix 熔断器相关注解，如 @HystrixCommand 等。

（3）Client
- OkHttpClient：使用 OKHTTP 库发起 HTTP 请求，提供了 feign.okhttp.OkHttpClient 类。
- RibbonClient：使用 Ribbon 库发起 HTTP 请求，提供了 feign.ribbo.RibbonClient 类。
- Apache HttpClient：使用 Apache 的 HttpClient 库发起 HTTP 请求，提供了 feign.httpclient.ApacheHttpClient 类。

5.2.2　Feign 整合 Hystrix

之前的示例中已经介绍可以通过增加 @HystrixCommand 注解方法增加熔断器，那么在 Feign 中可以在 @FeignClient 中配置 fallback 属性来开启熔断功能。例如：

```java
@FeignClient(name = "hello", fallback = HystrixClientFallback.class)
protected interface HystrixClient {
    @RequestMapping(method = RequestMethod.GET, value = "/hello")
    Hello iFailSometimes();
}
static class HystrixClientFallback implements HystrixClient {
    @Override
    public Hello iFailSometimes() {
        return new Hello("fallback");
    }
}
```

如果想在降级操作中处理抛出的异常，可以使用 @FeignClient 的 fallbackFactory 属性。例如：

```java
@FeignClient(name = "hello", fallbackFactory = HystrixClientFallbackFactory.class)
protected interface HystrixClient {
    @RequestMapping(method = RequestMethod.GET, value = "/hello")
    Hello iFailSometimes();
}
@Component
static class HystrixClientFallbackFactory implements FallbackFactory<HystrixClient> {
    @Override
    public HystrixClient create(Throwable cause) {
        return new HystrixClientWithFallBackFactory() {
            @Override
            public Hello iFailSometimes() {
                return new Hello("fallback; reason was: " + cause.getMessage());
            }
        };
    }
}
```

5.2.3 数据压缩

在对数据传输流量敏感的时候，可以考虑对 Feign 的 request 和 response 开启 GZIP 压缩处理。可以在配置文件中通过如下配置开启：

```
feign.compression.request.enabled=true
feign.compression.response.enabled=true
```

在 Web 服务中进行更详细的配置，指定 Media type 和压缩的最小值。

```
feign.compression.request.enabled=true
feign.compression.request.mime-types=text/xml,application/xml,application/json
feign.compression.request.min-request-size=2048
```

5.2.4 日志

对于创建的每个 Feign 客户端都会创建一个 Logger。默认情况下，使用客户端全类名作为日志名，Feign 日志只能设置为 DEBUG 级别。只需要在 application.yml 中添加如下配置即可：

```
logging.level.project.user.UserClient: DEBUG
```

可以为每个客户端配置一个 Logger.Level 对象，来告诉 Feign 如何输出日志。可以选择以下级别。

- NONE：无日志（默认）；
- BASIC：仅输出请求的方法、URL、response status code 及执行时间；
- HEADERS：带上 request 和 response 的 header 信息；
- FULL：包括 requset 和 response 的 header、body 及元数据。

例如：

```
@Configuration
public class FooConfiguration {
    @Bean
    Logger.Level feignLoggerLevel() {
        return Logger.Level.FULL;
    }
}
```

5.3 小结

Feign 大大简化了 HTTP 调用的开发代码量，提高了编码友好度，使得用户能够友好地面向接口开发，而不是生硬地面对 HTTP 参数进行开发。同时，Feign 完美地支持 Ribbon、Hystrix 等也是其天然的优势。接下来介绍 Hystrix 在熔断方面的本领。

第 6 章

熔断器：Hystrix

6.1 为什么要有熔断

微服务架构中，一般存在着很多服务单元。这样就有可能某个单元因为网络原因或自身问题而出现故障或延迟，导致调用方的对外服务也出现延迟。如果此时调用方的请求不断增加，时间一长就会由于等待故障方响应而形成任务积压，最终导致调用方自身服务的瘫痪。

为了解决这种问题，便出现了断路器（或称熔断器，Circuit Breaker）模式。断路器模式源于 Martin Fowler 的"Circuit Breaker"一文，我们日常生活中的断路器本身是一种开关装置，用于在电路上保护线路过载。当线路中有电器发生短路时，它能够及时切断故障电路，防止发生过载、发热甚至起火等严重后果。

微服务架构中的断路器的作用是：当某个服务单元发生故障（类似用电器短路）之后，通过断路器的故障监控（类似熔断保险丝），向调用方返回一个错误响应，而不是长时间的等待，这就不会使得线程被故障服务长时间占用而不释放，避免了故障在分布式系统中的蔓延。

假设服务 A 依赖服务 B 和服务 C，而 B 服务和 C 服务有可能继续依赖其他的服务，继续下去会使得调用链路过长，技术上称 1 → N 扇出，如图 6-1 所示。

如果在 A 的链路上某一个或几个被调用的子服务不可用或延迟较高，则会导致调用 A 服务的请求被堵住。堵住的请求会消耗、占用系统的线程、I/O 等资源，当该类请求越来越多，占用的计算资源越来越多的时候，会导致系统出现瓶颈，造成其他请求同样不可用，最终导致业务系统崩溃，又称"雪崩效应"。

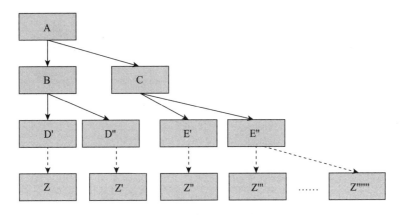

图 6-1 调用链路示例

对于正常情况下的访问,用户请求的多个服务(A、H、I、P)均能正常访问并返回,如图 6-2 所示。

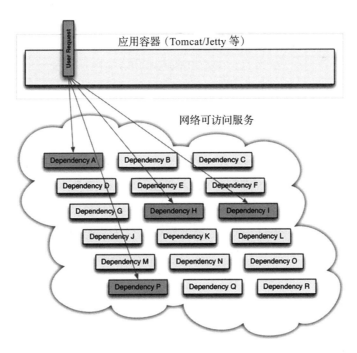

图 6-2 正常情况下的访问调用示意图

发生阻塞时的调用示意图如图 6-3 所示。

当请求的服务出现无法访问、异常、超时等问题时(图 6-3 中的 I),用户的请求将会被阻塞。雪崩发生时,如果多个用户的请求中都存在无法访问的服务,那么它们将陷入阻塞的状态。

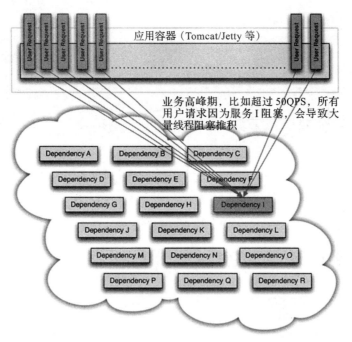

图 6-3　发生阻塞时的调用示意图

举例来说，一个汽车生产线生产不同的汽车，需要使用不同的零件，如果某个零件因为种种原因无法使用，就会造成整台车无法装配，陷入等待零件的状态，直到零件到位，才能继续组装。此时如果有很多个车型都需要这个零件，那么整个工厂都将陷入等待的状态，导致所有生产都陷入瘫痪。一个零件的波及范围不断扩大。

造成雪崩效应的原因如下：硬件故障、负载过大（如抢红包，双十一）、代码问题。

Hystrix 提供了熔断模式和隔离模式来解决或者缓解雪崩效应。这两种方案都属于阻塞发生之后的应对策略，而非预防性策略（如限流模式）。Hystrix 是在服务访问失败时降低阻塞的影响范围，避免整个服务被拖垮。

6.2　熔断原理

隔离模式：一个形象的解释是，对系统请求按类型划分成若干个的小岛，当某个小岛被火烧光了，不会影响到其他的小岛。

Hystrix 依赖的隔离架构如图 6-4 所示。

Hystrix 在用户请求和服务之间加入了线程池。Hystrix 为每个依赖调用分配一个小的线程池，如果线程池已满，调用将被立即拒绝，默认不采用排队，加速失败判定时间。

用户的请求将不再直接访问服务，而是通过线程池中的空闲线程来访问服务，如果线程池已满，则会进行降级处理，用户的请求不会被阻塞，至少可以看到一个执行结果（例如返回友好的提示信息），而不是无休止的等待或者看到系统崩溃。其本质是将服务视为资源，当请求该资源的数量超过了线程池中的数量限制时则不可以再对该资源进行访问，从而保护该资源不会过载而造成阻塞。

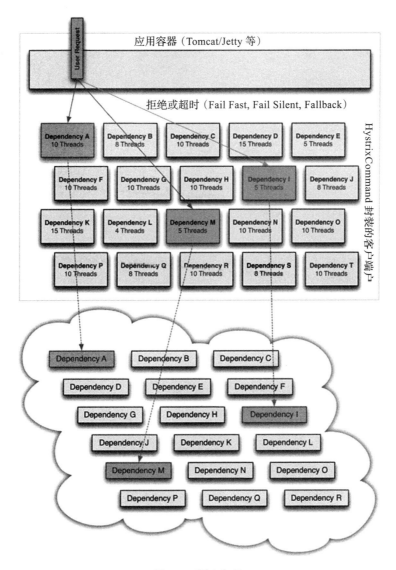

图 6-4　隔离架构

例如：将银行的柜员窗口看成服务，窗口外是排队办理业务的客户，假定每个客户的业务需要 5 分钟，则越靠后的用户等待时间越长。同时，排队的人越多，占用的空间越多（银行大厅被挤爆，导致想去别的窗口办理业务的人根本无法进入大厅，甚至队伍排到了大厅门外，此时已经雪崩）。线程池则规定了每个窗口前可以排队的人数上限（如 10 人），此时新来的第 11 人是不可以再到这个窗口前排队的，从而保证该窗口不会因为排队人数过多而阻塞。第 11 人获得的返回结果是去其他窗口办理，或者无法办理。这时大厅内还有足够的空间留给其他窗口办理业务的人员进出。A 窗口排满了，但是不影响 B 窗口的正常使用。这就是隔离的效果。如果不隔离，则会导致所有资源都被 A 窗口占用，其他窗口根本无法正常运作。下面看下几个关键的概念。

1. 熔断模式

该模式借鉴了电路熔断的理念，如果某个目标服务调用慢或者有大量超时，则熔断该服务的调用，后续请求不再调用该服务，直接返回并快速释放资源。如果目标服务情况好转，则恢复调用。

还是举之前银行柜员的例子，假定处理每个业务的时间是 5 分钟，当某个柜员的处理速度降低，超过了 5 分钟，或者干脆由于去吃午饭等原因根本不在座位上，此时熔断机制将不会允许再有客户到其窗口前排队（即使排队也无法正常办理业务，造成阻塞）。

下面来看看 Hystrix 是如何熔断的。

2. 熔断器

我们了解了 Hystrix 利用线程池实现了对服务的隔离。熔断器是位于线程池之前的组件。用户请求某一服务之后，Hystrix 会先经过熔断器，此时如果熔断器的状态是打开（跳起），则说明已经熔断，将直接进行降级处理，不会继续将请求发到线程池。

套用银行柜员的例子，柜员相当于服务，窗口前排队的是线程池，大堂经理则可以看成熔断器。通常的流程是：客户进门，告诉大堂经理要办什么业务，这时大堂经理会判断客户请求的窗口是否能正常处理业务，如果正常，他就会让客户到该窗口排队（也就是进入了线程池），如果不正常，他根本不会让客户去排队。

熔断器相当于在线程池之前的一层屏障。

下面来看一下熔断器的工作原理，如图 6-5 所示。

每个熔断器默认维护 10 个 bucket，每秒创建一个 bucket，每个 bucket 记录成功、失败、超时、拒绝的次数，当有新的 bucket 被创建时，最旧的 bucket 会被抛弃。

3. 熔断算法

判断是否进行熔断的依据是：根据 bucket 中记录的次数，计算错误率。当错误率超过预设的值（默认是 50%）且 10s 内超过 20 个请求，则开启熔断。

4. 熔断恢复

被熔断的请求并不是永久被切断，而是暂停一段时间（默认是 5s）之后，允许部分请求通过，若请求都是健康的（ResponseTime<250ms），则对请求进行健康恢复（取消熔断），如果是不健康的，则继续熔断。

服务调用的各种结果（成功、异常、超时、拒绝）都会上报给熔断器，计入 bucket 参与计算。

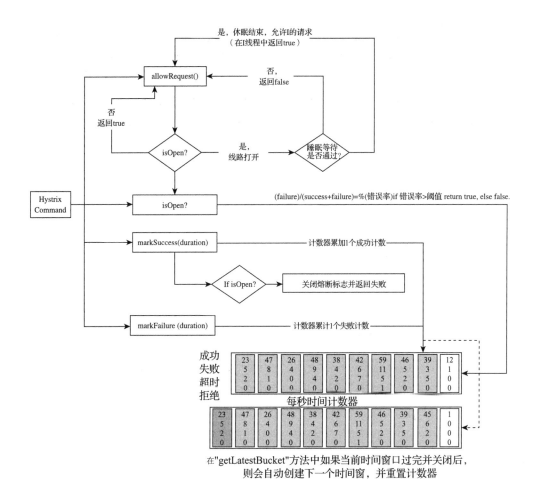

图 6-5 熔断器的工作原理

6.3 使用 Hystrix

接下来尝试模拟一个调用方，并进行熔断配置。

在 Consumer 服务消费方的 Maven 中引入 Hystrix 的依赖。

```
<dependency>
    <groupId>org.springframework.cloud</groupId>
    <artifactId>spring-cloud-starter-hystrix</artifactId>
</dependency>
```

接着对上文中的 FeignClientApplication.java 进行改造。增加 @EnableCircuitBreaker 注解，其实 @EnableHystrix 也可以。不过 @EnableCircuitBreaker 是 Spring Cloud 对 Netflix 捐赠的 Hystrix 进行抽象后新增的注解，所以还是推荐使用 @EnableCircuitBreaker。

在 sayHello 方法上增加 @HystrixCommand，表示对该方法开启熔断检测功能，一旦发生异常，则返回 fallbackMethod 中指定的方法提供的 Mock 参数。当我们为方法增加这个注解后，Spring 则会自动创建一个熔断代理器，负责记录失败次数、频率等相关信息，然后根据配置要求，进行熔断的开关。

在 sayHello 方法上通过随机数模拟调用服务失败的频率，让其有一半的几率调用失败。

```
@SpringBootApplication
@EnableFeignClients
@EnableEurekaClient
@Slf4j
@RestController
@EnableCircuitBreaker
public class FeignClientApplication {
    @Autowired
    private IBizApi bizApi;
    @RequestMapping("/sayHello")
    @HystrixCommand(fallbackMethod = "faildSay")
    public String sayHello(){
        Random random = new Random();
        if (random.nextInt(10)<5){
            throw new RuntimeException("myexception");
        }
        return  bizApi.sayHello("haha");
    }
    public String faildSay(){
        return "don't say";
    }
    public static void main(String[] args) {
        SpringApplication.run(FeignClientApplication.class, args);
    }
}
```

接下来尝试请求的 http://localhost:8081/sayHello，会发现有一半的几率返回的是熔断方法提供的 Mock 返回值 don't say。

可以通过以下几种方式对 Hystrix 进行配置。

1. 注解直接配置

Hystrix 提供了非常方便的注解配置方式，使用 @HystrixCommand 注解并对 command-Properties 设置配置项即可进行自行配置，代码如下：

```
@HystrixCommand(commandProperties = {
        @HystrixProperty(name = "execution.isolation.thread.timeoutInMilliseconds",
            value = "500")
    })
    public User getUserById(String id) {
        return userResource.getUserById(id);
    }
```

ConfigurationManager 配置：

```
ConfigurationManager.getConfigInstance().setProperty("hystrix.command.
    getUserById.execution.isolation.thread.timeoutInMilliseconds", "500");
```

线程池相关配置:

```
@HystrixCommand(commandProperties = {
        @HystrixProperty(name = "execution.isolation.thread.timeoutIn-
            Milliseconds", value = "500")
    },
        threadPoolProperties = {
            @HystrixProperty(name = "coreSize", value = "30"),
            @HystrixProperty(name = "maxQueueSize", value = "101"),
            @HystrixProperty(name = "keepAliveTimeMinutes", value = "2"),
            @HystrixProperty(name = "queueSizeRejectionThreshold", value = "15"),
            @HystrixProperty(name = "metrics.rollingStats.numBuckets",
                value = "12"),
            @HystrixProperty(name = "metrics.rollingStats.timeInMilliseconds",
                value = "1440")
        })
    public User getUserById(String id) {
        return userResource.getUserById(id);
    }
```

2. 配置文件式

同其他项目一样，能够使用注解配置的当然也能够使用配置文件进行配置。配置项示例如下：

```
hystrix:
    command:
        <commandKey>:   #通过HystrixCommand的commandKey属性指定
            execution:
                isolation:
                    thread:
                        timeoutInMilliseconds: 5000
```

Hystrix 的大部分配置都以 hystrix.command.[HystrixCommandKey] 开头。其中 [HystrixCommandKey] 是可变的，默认是 default，即 hystrix.command.default（对于 Zuul 而言，CommandKey 就是 service ID）。

常见的有以下几个配置。

- hystrix.command.default.execution.isolation.thread.timeoutInMilliseconds：用来设置 thread 和 semaphore（计数信号量）两种隔离策略的超时时间，默认值是 1000。
- hystrix.command.default.execution.isolation.semaphore.maxConcurrentRequests：设置当使用 ExecutionIsolationStrategy.SEMAPHORE 时，HystrixCommand.run() 方法允许的最大请求数。如果达到最大并发数时，后续请求会被拒绝。
- hystrix.command.default.execution.timeout.enabled：是否开启超时，默认是 true。
- hystrix.command.default.execution.isolation.thread.interruptOnTimeout：发生超时是否中断线程，默认是 true。

- hystrix.command.default.execution.isolation.thread.interruptOnCancel：取消时是否中断线程，默认是 false。
- hystrix.command.default.circuitBreaker.requestVolumeThreshold：当在配置时间窗口内达到此数量的失败后，进行短路，默认是 20 个。
- hystrix.command.default.circuitBreaker.sleepWindowInMilliseconds：短路多久以后开始尝试恢复，默认 5s。
- hystrix.command.default.circuitBreaker.errorThresholdPercentage：出错百分比阈值，当达到此阈值后，开始短路，默认 50%。
- hystrix.command.default.fallback.isolation.semaphore.maxConcurrentRequests：调用线程允许请求 HystrixCommand.GetFallback() 的最大数量，默认 10，超出时将会有异常抛出。注意：该项配置对于 thread 隔离模式也起作用。

注意 以上就是列举的一些常见配置，更多内容可参考 https://github.com/Netflix/Hystrix/wiki/Configuration。

6.4 Hystrix 数据监控

当熔断发生的时候需要迅速响应解决问题，避免故障进一步扩散，那么对熔断的监控就变得非常重要。熔断的监控现在有两款工具：**Hystrix-dashboard** 和 Turbine。

Hystrix Dashboard 是一款针对 Hystrix 进行实时监控的工具，通过 Hystrix Dashboard 可以直观地看到各 Hystrix 命令的请求响应时间、请求成功率等数据。但是若只使用 Hystrix Dashboard，则只能看到单个应用内的服务信息，这明显不够。我们需要一个工具能汇总系统内多个服务的数据并显示到 Hystrix Dashboard 上，这个工具就是 Turbine。

6.4.1 健康指示器

同样也可以使用 Spring Boot 通用的健康指示器 /health 来对应用进行健康度检查。如下为健康指示器的示例返回值：

```
{
    "hystrix": {
        "openCircuitBreakers": [
            "StoreIntegration::getStoresByLocationLink"
        ],
        "status": "CIRCUIT_OPEN"
    },
    "status": "UP"
}
```

当开启 health 的健康端点时，能够查到应用健康信息是一个汇总的信息，如果我们获取到的信息是 {"status":"UP"}，则表示服务正常运行。表 6-1 是内置健康状态类型对应的

HTTP 状态码列表。

表 6-1 健康指示器响应码列表

状态码	对应 HTTP 响应
DOWN	SERVICE_UNAVAILABLE (503)
OUT_OF_SERVICE	SERVICE_UNAVAILABLE (503)
UP	HTTP_SUCCESS(200)
UNKNOWN	UNKNWN(200)

6.4.2 监控面板

在需要被监测的项目中，开启监控需要引入依赖 spring-boot-starter-actuator。这会提供一个 /hystrix.stream 管理接口。

```
<dependency>
    <groupId>org.springframework.boot</groupId>
    <artifactId>spring-boot-starter-actuator</artifactId>
</dependency>
```

请求 http://localhost:8081//hystrix.stream，可以看到不断刷新的数据，一直在实时表示着需要被监控的一些参数项的刷新。

Hystrix 的一个主要的特性就是每一条 Hystrix Command 操作的各个方面都可以度量检测。Hystrix 的监控面板页面可以有效地显示每一个断路器的健康状态，如图 6-6 所示。

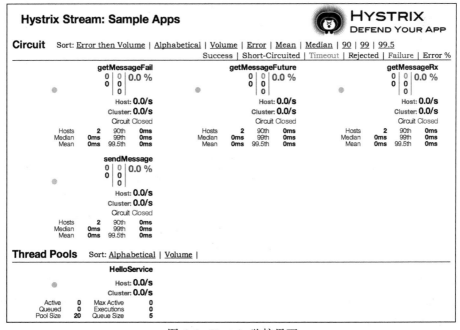

图 6-6 Hystrix 监控界面

要创建仪表盘应用，可以首先新建一个 Maven 项目并在 pom 中引入仪表盘的 Maven 依赖：

```
<dependency>
    <groupId>org.springframework.cloud</groupId>
    <artifactId>spring-cloud-starter-hystrix-dashboard</artifactId>
</dependency>
```

在主类通过 @EnableHystrixDashboard 注解来开启仪表板：

```
@SpringBootApplication
@EnableHystrixDashboard
public class MydashboardApplication {
    public static void main(String[] args) {
        SpringApplication.run(MydashboardApplication.class, args);
    }
}
```

在配置文件 application.yml 中指定端口：

```
server:
    port: 8412
```

然后启动项目，就可以通过 http://localhost: 8412/hystrix 访问仪表板页面，界面如图 6-7 所示。

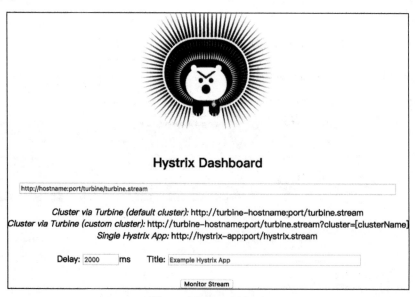

图 6-7　监控面板主页

仪表板相关数据都是通过 /hystrix.stream 接口提供的，所以需要在正中间的文本框填入 http://localhost:8081/hystrix.stream。下面的 Delay 文本框表示多久刷新一次数据，Title 为页面显示的标题，如图 6-8 所示。

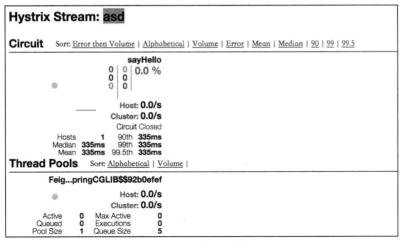

图 6-8　指标示例

6.4.3　聚合监控

在实际的生产环境中可能有许多服务，每个服务有许多实例，不可能同时打开很多个仪表盘。我们需要一个聚合的监控面板。Turbine 提供了一个仪表板页面来聚合所有 /hystrix.stream 相关数据，同时把所有数据合成到一个 /turbine.stream 接口中。聚合数据架构如图 6-9 所示。

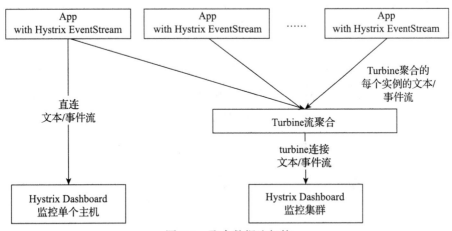

图 6-9　聚合数据流架构

在上文的 Hystrix Dashboard 项目中继续增加 Turbine 的依赖。

```
<dependency>
    <groupId>org.springframework.cloud</groupId>
    <artifactId>spring-cloud-starter-turbine</artifactId>
</dependency>
```

配置文件 application.yml：

```yaml
server:
    port: 8412
eureka:
    client:
        serviceUrl:
            defaultZone: http://localhost:8989/eureka/
turbine:
    aggregator:
        clusterConfig: MAIN                              #集群名字，可以多个
    appConfig: eurekaclient                              #turbine监控的服务，可以有多个
    clusterNameExpression: metadata['cluster']
```

我们在文件中依旧需要指定 Eureka 的地址，用于使 Turbine 能够去注册中心查找需要监测的服务实例。

同样，在被监控项目（如服务消费者）中也需要进行对应配置，需要保证 Turbine 中的 clusterConfig 的配置与如下配置值保持一致，并且需要大写：

```yaml
eureka:
    instance:
        metadata-map:
            cluster: MAIN
```

接下来就可以直接请求 http://localhost:8412/turbine.stream?cluster=MAIN。

如果我们启动多个消费者，可以观察到原本 /hystrix.stream 的数据被聚合到了一起，将 Turbine 的聚合地址填入监控面板页面，即可看到聚合之后的监控图形化信息，如图 6-10 所示。

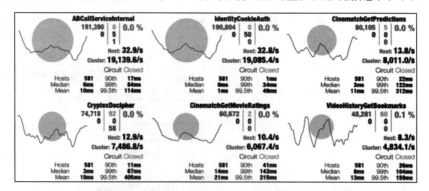

图 6-10　Turbine 聚合 Hystrix 的图形化显示

6.5　小结

Hystrix 之于分布式就像大坝之于洪川，当企业级服务达到一定量级时一定缺不了 Hystrix 这种能够灵活配置、稳定的熔断组件。同时其监控仪表盘也非常详尽、友好，以至不用二次开发或者稍加二次开发即可投入使用。在介绍完 Hystrix 之后，我们将介绍另外一个重量级组件 Zuul。

第 7 章 Chapter 7

路由网关：Zuul

在微服务架构模式下，后端服务的实例数一般是动态的，对于客户端而言很难发现动态改变的服务实例的访问地址信息。因此，在基于微服务的项目中为了简化前端的调用逻辑，通常会引入 API Gateway 作为轻量级网关，同时 API Gateway 中也会实现相关的认证逻辑，从而简化内部服务之间相互调用的复杂度。

Spring Cloud 体系中支持 API Gateway 落地的技术就是 Zuul。Spring Cloud Zuul 路由是微服务架构中不可或缺的一部分，提供动态路由、监控、弹性和安全等边缘服务。

Zuul 是 Netflix 出品的一个基于 JVM 路由和服务端的负载均衡器。它的具体作用就是服务转发，接收并转发所有内外部的客户端调用。Zuul 可以作为资源的统一访问入口，同时也可以在网关做一些权限校验等类似的功能。同时它也集成了 Eureka、Ribbon、Hystrix 等，代理服务会使用 Ribbon 的负载来转发请求，所有的请求也将由 Hystrix 命令来执行，这样就自带降级、断路器等功能，简单架构如图 7-1 所示。

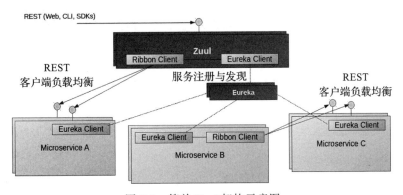

图 7-1　简单 Zuul 架构示意图

7.1 使用 Zuul

先尝试运行一个基于 Zuul 的网关来对接入请求进行转发，以便对 Zuul 有个直观感受。首先在 Maven 的 pom.xml 中增加如下依赖：

```xml
<dependency>
    <groupId>org.springframework.cloud</groupId>
    <artifactId>spring-cloud-starter-zuul</artifactId>
</dependency>
<dependency>
    <groupId>org.springframework.cloud</groupId>
    <artifactId>spring-cloud-starter-eureka</artifactId>
</dependency>
```

我们在启动类上标注 @EnableZuulProxy 注解。

```
@Spring CloudApplication
@EnableZuulProxy
public class MyZuulApplication {
    public static void main(String[] args) {
        SpringApplication.run(MyZuulApplication.class, args);
    }
}
```

这里可以看到一个新的注解 @Spring CloudApplication，此注解包含了一个 Spring Cloud 项目的常用注解 @SpringBootApplication、@EnableDiscoveryClient、@EnableCircuitBreaker。在后文需要的场景我们也会直接标注此注解。

```
@Target(ElementType.TYPE)
@Retention(RetentionPolicy.RUNTIME)
@Documented
@Inherited
@SpringBootApplication
@EnableDiscoveryClient
@EnableCircuitBreaker
public @interface Spring CloudApplication {}
```

我们在 application.yml 中增加如下配置，配置三个路由策略分别转发到某个服务或者特定 URL：

```yaml
zuul:
    ignored-services: "*"
    routes:
        myprovider:
            path: /myproviderUrl/**
            serviceId: myprovider
        myurl:
            path: /myurl/**
            url: http://localhost:8085/
```

```
            myurl2:
                path: /myurl2/**
                serviceId: myprovider
server:
    port: 9876
eureka:
    client:
        serviceUrl:
            defaultZone: http://localhost:8989/eureka/
```

在 routes 节点下，可以以三种方式来配置 Zuul 的路由规则。

```
#第一种
<serviceId>:                              #对应Eureka中的ServiceId，规则名与ServiceId相同
    path: /myproviderUrl/**               #将哪些URL转发
#第二种
myurl:                                    #自定义规则名称
    path: /myurl/**                       #将哪些URL转发
    url: http://localhost:8085/           #转发到哪个URL，如果只使用此路由规则，可以不依赖Eureka
myurl2:                                   #自定义规则名称
    path: /myurl2/**                      #将哪些URL转发
    serviceId: myprovider                 #对应Eureka中的ServiceId
```

启动主类，然后尝试访问：

- http://localhost:9876/myproviderUrl/sayHello?name=test；
- http://localhost:9876/myurl/sayHello?name=test；
- http://localhost:9876/myurl2/sayHello?name=test。

可以看到均能正常访问，并在服务提供方 Provider 中也能看到被请求的日志。

7.2 业务场景深入解析

上文运行的简易示例当然是不适用企业级应用场景的，需要对其进行深入剖析并自行调优以适应实际业务场景。

1. 过滤器

除了请求的转发，可能要有一些业务场景，比如接口权限校验、限流、统计等。那也可以通过 Zuul 的过滤器来实现。

Zuul 的大部分功能都是通过过滤器来实现的。Zuul 中定义了 4 种标准过滤器类型，这些过滤器类型对应于请求的典型生命周期，可以借助 Spring 的 Aop 实现便于理解，如图 7-2 所示。

下面分别解释生命周期的每个阶段对应的职责。

- PRE：这种过滤器在请求被路由之前调用。我们可利用这种过滤器实现身份验证、在集群中选择请求的微服务、记录调试信息等。

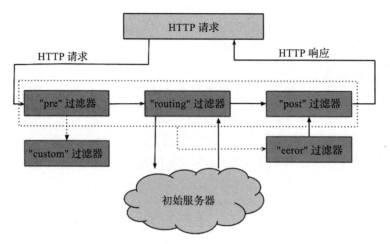

图 7-2　Zuul 生命周期

- ROUTING：这种过滤器将请求路由到微服务。这种过滤器用于构建发送给微服务的请求，并使用 Apache HttpClient 或 Netfilx?Ribbon 请求微服务。
- POST：这种过滤器在路由到微服务以后执行。这种过滤器可用来为响应添加标准的 HTTP Header、收集统计信息和指标、将响应从微服务发送给客户端等。
- ERROR：在其他阶段发生错误时执行该过滤器。

除了默认的过滤器类型外，Zuul 还允许用户创建自定义的过滤器类型。例如，可以定制一种 STATIC 类型的过滤器，直接在 Zuul 中生成响应，而不将请求转发到后端的微服务。可以通过继承 com.netflix.zuul.ZuulFilter 并实现其中的抽象方法来达到所需要的目的。

接下来，自定义一个 ZuulFilter：

```
public class MyFilter extends ZuulFilter {
    @Override
    public String filterType() {
        return "pre";
    }
    @Override
    public int filterOrder() {
        return 0;
    }
    @Override
    public boolean shouldFilter() {
        return true;
    }
    @Override
    public Object run() {
        RequestContext currentContext = RequestContext.getCurrentContext();;
        HttpServletRequest request = currentContext.getRequest();
        HttpServletResponse response = currentContext.getResponse();
```

```
        return null;
    }
}
```

可以看到自定义的 MyFilter，并重写了一些 ZuulFilter 中的方法实现。下面分别就这几个方法对应的含义做一下解析。

- filterType：返回过滤器的类型。有 pre、route、post、error 等取值，分别对应上文的几种过滤器。详细内容可以参考 com.netflix.zuul.ZuulFilter.filterType()? 中的注释。
- filterOrder：返回一个 int 值来指定过滤器的执行顺序，不同的过滤器允许返回相同的数字。
- shouldFilter：返回一个布尔值来判断该过滤器是否要执行，true 表示执行，false 表示不执行。
- run：过滤器的具体逻辑。可以通过 RequestContext 得到 J2EE 的 Request、Response 类来实现需要的逻辑。返回值在目前版本中暂时不起作用。

最后，将定义好的过滤器定义成 Spring 的 Bean 即可：

```
@Bean
public ZuulFilter myFilter(){
    return new MyFilter();
}
```

对于启用/禁用过滤器，当然不愿意每次去修改过滤器源代码。可以通过配置文件对过滤器进行控制，在 application.yml 中增加如下配置即可：

```
zuul:
    <SimpleClassName>:
        <filterType>:
            disable: true
```

比如，禁用刚才自定义的 MyFilter：

```
zuul:
    MyFilter:
        pre:
            disable: true alse
```

2. 正则表达式匹配

有时候我们的匹配接收来的请求可能不是固定的，所以希望通过如正则表达式等方法来进行匹配，Zuul 同样提供了这样的支持。

可以定义一个由 Spring 管理的 PatternServiceRouteMapper。通过构造函数传入 servicePattern、routePattern，当请求路径为 "/v1/myusers/" 时，Zuul 会将其映射到 "myusers-v1" 服务上，代码如下：

```
@Bean
public PatternServiceRouteMapper serviceRouteMapper() {
```

```
    String servicePattern = "(?<name>^.+)-(?<version>v.+$)";
    String routePattern = "${version}/${name}";
    return new PatternServiceRouteMapper(servicePattern,routePattern);
}
```

3. 统一前缀

可以通过 zuul.prefix 为所有的映射增加统一的前缀，如 /api。默认情况下，代理会在转发前去掉这个前缀。如果需要转发时带上前缀，可以配置 zuul.stripPrefix=false 来关闭这个默认行为。通过 application.yml 增加此配置即可，例如：

```
zuul:
    routes:
        users:
            path: /myusers/**
            stripPrefix: false
```

 zuul.stripPrefix 只会对 zuul.prefix 的前缀起作用，对于 path 指定的前缀不会起作用。

4. 请求重试

当 Zuul 转发请求失败时，可以通过配置使 Ribbon 客户端进行重试，默认值为 False。

```
zuul:
retryable: false
```

5. Http-Header 信息转发

默认情况下，对于转发的请求，头信息会添加 X-Forwarded-Host 信息。可以通过配置 zuul.addProxyHeaders = false 来关闭这个特性。前缀默认会被剥离，但是后端可以通过 X-Forwarded-Prefix 属性获取前缀。

6. 默认根路径

如果配置了一个默认路由 "/"，那么当应用带有 @EnableZuulProxy 注解时，可以得到一个独立的服务。例如：对于 "zuul.route.home:/"，所有的请求（"/**"）都会转发到 "home" 服务。

7. 忽略匹配

ignoredPatterns 可以忽略 URL。要注意匹配是从前缀开始的，并且会匹配所有符合条件的服务。这个示例中，"/myusers/101" 将会转发到 "/101" 到 "users" 服务。但是如果包含 "/admin" 则不会转发。通过 application.yml 增加此配置即可，例如：

```
zuul:
    ignoredPatterns: /**/admin/**
    routes:
        users: /myusers/**
```

8. 匹配顺序

如果想按配置的顺序进行路由规则控制，则需要使用 YAML，如果是使用 properties 文件，则会丢失顺序。通过 application.yml 增加此配置即可，例如：

```
zuul:
    routes:
        users:
            path: /myusers/**
        legacy:
            path: /**
```

如果使用 properties 文件进行配置，则 legacy 可能先生效，这样 users 就没有效果了。

9. 设置 Zuul Http Client

Zuul 默认会使用 Apache HTTP 客户端，而不是使用 Ribbon 的 Rest Client。如果想要使用 REST Client 或者 okhttp3.OkHttpClient，可以配置 ribbon.restclient.enabled=true 或 ribbon.okhttp.enabled=true。

10. 敏感 Header 过滤

在请求的转发中默认会转发 HTTP 的 Header 信息，然而可能有些敏感信息不能被转发给下游系统，如 Cookie。可以通过 sensitiveHeaders 来进行配置，各项之间用逗号分隔。通过 application.yml 增加此配置即可，例如：

```
zuul:
    routes:
        users:
            path: /myusers/**
            sensitiveHeaders: Cookie,Set-Cookie,Authorization
            url: https://downstream
```

11. 路由信息端点

当我们使用了 @EnableZuulProxy 时，默认会开启一个额外的 HTTP 接口 "/routes"。GET 请求这个接口会返回路由映射的清单。POST 请求可以强制刷新某个存在的路由。

当 /routes 被权限校验拦截下来，可以通过如下配置暂时关闭：

```
management:
    security:
        enabled: false
```

请求 http://localhost:9876/routes，则会看到如下返回信息：

```
{"/myproviderUrl/**":"myprovider","/myurl/**":"http://localhost:8085/","/myurl2/**":"myprovider"}
```

注意　服务信息变更后，路由映射会自动更新。通过这个接口强制立即刷新。

12. 转发支持

比如，做系统迁移、提供升级的过程中，可能需要一个灰度阶段进行过渡。在这一点上 Zuul 代理实现也比较方便，通过 forward 关键字，可以让客户端的老接口全部走代理，通过路由转发到新的系统接口上。通过 application.yml 增加此配置即可，例如：

```yaml
zuul:
    routes:
        newSys:
            path: /newSys/**
            url: http://newSys.example.com
        oldSys:
            path: /oldSys/**
            url: forward:/newSys
```

13. 文件上传

如果已经使用 @EnableZuulProxy，就可以通过代理进行文件上传操作，当然文件不能过大。大文件最好直接通过 Spring MVC 的 DispatcherServlet 来处理 /zuul/* 请求。例如：假设配置了 zuul.routes.customers=/customers/**，就可以用 POST 请求 /zuul/customers/* 来处理大文件的上传操作。Servlet 路径可以通过 zuul.servletPath 来指定。

如果文件确实比较大，那么会导致上传过程中处理超时。所以，还需要配置一下超时设置。通过 application.yml 增加此配置即可，例如：

```yaml
hystrix.command.default.execution.isolation.thread.timeoutInMilliseconds: 60000
ribbon:
    ConnectTimeout: 3000
    ReadTimeout: 60000
```

14. 服务降级

服务降级可以通过创建一个 Bean——ZuulFallbackProvider 来为 Zuul 调用链路实现。在这个 Bean 中需要指定一个路由的 ID，当触发降级时，由这个路由来返回 ClientHttpResponse。下面展示了一个简单的 ZuulFallbackProvider 实例：

```java
class MyFallbackProvider implements ZuulFallbackProvider {
    @Override
    public String getRoute() {
        return "myprovider";
    }
    @Override
    public ClientHttpResponse fallbackResponse() {
        return new ClientHttpResponse() {
            @Override
            public HttpStatus getStatusCode() throws IOException {
                return HttpStatus.OK;
            }
            @Override
```

```java
            public int getRawStatusCode() throws IOException {
                return 200;
            }
            @Override
            public String getStatusText() throws IOException {
                return "OK";
            }
            @Override
            public void close() {
            }
            @Override
            public InputStream getBody() throws IOException {
                return new ByteArrayInputStream("fallback".getBytes());
            }
            @Override
            public HttpHeaders getHeaders() {
                HttpHeaders headers = new HttpHeaders();
                headers.setContentType(MediaType.APPLICATION_JSON);
                return headers;
            }
        };
    }
}
```

7.3 小结

市面上的分发网关非常多，但是能够像 Zuul 这样经历过各大互联网公司大流量检验同时功能又如此完善的网关产品并不多。另外，与 Spring Cloud 生态的完美融合让其成为微服务架构中不可或缺的一个关键节点。

第 8 章

网关新选择：Gateway

Spring Cloud Gateway 是 Spring 官方基于 Spring 5.0、Spring Boot 2.0 和 Project Reactor 等技术开发的网关，Spring Cloud Gateway 旨在为微服务架构提供一种简单而有效的统一的 API 路由管理方式。Spring Cloud Gateway 作为 Spring Cloud 生态系中的网关，目标是替代 Netflix Zuul。其不仅提供统一的路由方式，并且基于 Filter 链的方式提供了网关基本的功能，例如安全、监控/埋点和限流等。

Spring Cloud Gateway 具有如下特征：

- 基于 Java 8、Spring Framework 5 和 Spring Boot 2 开发；
- 支持动态路由；
- 内置到 Spring Handler 映射中的路由匹配；
- 基于 HTTP 请求的路由匹配；
- 过滤器基于匹配的路由机制；
- 过滤器可以修改下游 HTTP 请求和 HTTP 响应；
- 通过 API 或配置驱动；
- 支持 Spring Cloud DiscoveryClient 配置路由，与服务发现与注册配合使用。

1. 与 Zuul 对比

Zuul 基于 servlet 2.5（使用 3.x），使用阻塞 API。它不支持任何长连接，如 Web Sockets。而 Gateway 建立在 Spring Framework 5、Project Reactor 和 Spring Boot 2 之上，使用非阻塞 API，支持 Web Sockets，并且由于它与 Spring 紧密集成，所以将会有更好的开发体验。

2. 术语解释

在 Spring Cloud Gateway 中包含了如下几个概念，我们先初步了解一下。

- Route：路由网关的基本模块。它由 ID、目标 URI、断言集合和过滤器集合定义。如果聚合断言组为真，则匹配路由。
- Predicate：实际上就是 Java 8 Function Predicate 的断言功能。输入类型是 Spring Framework ServerWebExchange。该功能允许开发人员自行匹配来自 HTTP 请求的任何内容，例如 HTTP 头或参数。
- Filter：过滤器，Spring Framework GatewayFilter 实例，可以在发送下游请求之前或之后修改请求和响应。

3. 工作原理

当客户端向 Spring Cloud Gateway 发出请求时，如果网关处理程序判断请求与路由规则相匹配，则将请求继续下发到网关 Web 处理节点，由 Web 处理节点将其下发到过滤器链。过滤器被虚线划分的原因是过滤器可以在发送代理请求之前或之后执行逻辑。执行所有 PRE 阶段过滤器逻辑，然后进行代理请求发送到真正的业务处理模块。在发出代理请求之后，执行 POST 过滤器逻辑对响应结果进行过滤，如图 8-1 所示。

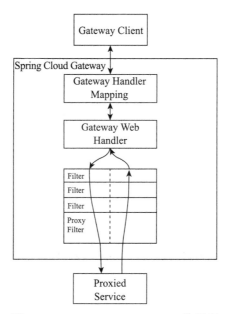

图 8-1　Spring Cloud Gateway 工作原理

8.1　使用 Gateway

首先示例启动两个服务：Gateway-Server 和 User-Server。模拟的场景是，客户端请求后端服务，网关提供后端服务的统一入口。后端的服务都注册到服务发现 Consul，网关通过负载均衡转发到具体的后端服务。

1. 用户服务

我们模拟一个用户服务，并将服务注册到 Consul 上，且该服务提供一个接口 /test。
先在项目中 Maven 的 pom 文件中添加依赖如下：

```xml
<dependency>
    <groupId>org.springframework.cloud</groupId>
    <artifactId>spring-cloud-starter-consul-discovery</artifactId>
</dependency>

<dependency>
    <groupId>org.springframework.boot</groupId>
    <artifactId>spring-boot-starter-web</artifactId>
</dependency>
```

在项目的配置文件中添加如下配置项：

```yaml
spring:
    application:
        name: user-service
    cloud:
        consul:
            host: 192.168.1.204
            port: 8500
            discovery:
                ip-address: ${HOST_ADDRESS:localhost}
                port: ${SERVER_PORT:${server.port}}
                healthCheckPath: /health
                healthCheckInterval: 15s
                instance-id: user-${server.port}
                service-name: user
server:
    port: 8005
management:
    security:
        enabled: false
```

由于上述配置项等都在前文中介绍过，在此不再解释。
我们编写一个 Spring Boot 启动类，并暴露 /test 接口：

```java
@SpringBootApplication
@RestController
@EnableDiscoveryClient
public class GatewayUserApplication {

    public static void main(String[] args) {
        SpringApplication.run(GatewayUserApplication.class, args);
    }

    @GetMapping("/test")
    public String test() {
```

```
            return "ok";
    }
}
```

至此，一个简单的业务应用就完成了，我们可以自行测试访问暴露 /test 接口，返回 ok 即可。

2. 网关服务

网关服务提供多种路由配置、路由断言工厂和过滤器工厂等功能。我们可以一起来逐一测试。

首先先在 Maven 的 pom 文件中添加需要引入的依赖：

```xml
<dependency>
    <groupId>org.springframework.boot</groupId>
    <artifactId>spring-boot-actuator</artifactId>
</dependency>
//依赖于WebFlux，必须引入
<dependency>
    <groupId>org.springframework.boot</groupId>
    <artifactId>spring-boot-starter-webflux</artifactId>
</dependency>
<dependency>
    <groupId>org.springframework.cloud</groupId>
    <artifactId>spring-cloud-gateway-core</artifactId>
</dependency>
//服务发现组件，排除Web依赖
<dependency>
    <groupId>org.springframework.cloud</groupId>
    <artifactId>spring-cloud-starter-consul-discovery</artifactId>
    <version>2.0.0.M6</version>
    <exclusions>
        <exclusion>
            <groupId>org.springframework.boot</groupId>
            <artifactId>spring-boot-starter-web</artifactId>
        </exclusion>
    </exclusions>
</dependency>
//Kotlin依赖
<dependency>
    <groupId>org.jetbrains.kotlin</groupId>
    <artifactId>kotlin-stdlib</artifactId>
    <version>${kotlin.version}</version>
    <optional>true</optional>
</dependency>
<dependency>
    <groupId>org.jetbrains.kotlin</groupId>
    <artifactId>kotlin-reflect</artifactId>
    <version>${kotlin.version}</version>
    <optional>true</optional>
</dependency>
```

> **注意**：如上引入了 Kotlin 相关的依赖，这里需要支持 Kotlin 的路由配置。Spring Cloud Gateway 的使用需要排除 Web 相关的配置，引入的是 WebFlux 的引用，应用启动时会检查，必须引入。

然后尝试配置网关的服务路由，即根据请求中的 serviceId 将请求匹配并转发到其所对应的服务实例。由于在前面的配置已经引入了相应的依赖，那么下面开始编写路由配置代码：

```
@Bean
public RouteDefinitionLocator discoveryClientRouteDefinitionLocator(DiscoveryCli
    ent discoveryClient) {
    return new DiscoveryClientRouteDefinitionLocator(discoveryClient);
}
```

将 DiscoveryClient 注入到 DiscoveryClientRouteDefinitionLocator 的构造函数中，然后将 application.yml 配置文件中新增如下配置项：

```
spring:
  cloud:
    gateway:
      locator:
        enabled: true
      default-filters:
      - AddResponseHeader=X-Response-Default-Foo, Default-Bar
      routes:
      # ====================================
      - id: service_to_user
        uri: lb://user
        order: 8000
        predicates:
        - Path=/user/**
        filters:
        - StripPrefix=1
```

上面的配置开启了 DiscoveryClient 定位器的实现。路由定义了，所有请求路径以 /user 开头的请求，都将会转发到 user 服务，并应用路径的过滤器截取掉路径的第一部分前缀，即访问 /user/test 的实际请求转换成了 lb://user/test。

至此，一个最简单的场景已经搭建完毕，接下来将全面了解 Gateway 路由断言、过滤器等功能。

8.2 路由断言

Spring Cloud Gateway 将路由作为 Spring WebFlux HandlerMapping 基础结构的一部分进行匹配。Spring Cloud Gateway 包含许多内置的路由断言（Route Predicate）。所有这些断

言都可以用来匹配 HTTP 请求的不同属性。多路断言工厂可以组合并通过逻辑连接，如与或非等进行连接来实现。我们下面将逐一介绍。

1. After 路由断言

After 路由断言只有一个参数：日期时间。此断言用以匹配在当前日期时间之后发生的请求。

application.yml 配置示例如下：

```
spring:
    cloud:
        gateway:
            routes:
            - id: after_route
              uri: http://example.org
              predicates:
              - After=2019-01-20T17:42:47.789-07:00[America/Denver]
```

此断言匹配任意晚于 2019 年 1 月 20 日 17:42 Mountain Time（Denver）这个时间点的所有请求。

2. Before 路由断言

与 After 路由断言类似，Before 路由断言同样也采用一个参数：日期时间。此断言匹配在当前日期时间之前发生的请求。

application.yml 配置如下：

```
spring:
    cloud:
        gateway:
            routes:
            - id: before_route
              uri: http://example.org
              predicates:
              - Before=2019-01-20T17:42:47.789-07:00[America/Denver]
```

此断言匹配任意早于 2019 年 1 月 20 日 17:42 Mountain Time（Denver）的任何请求。

3. Between 路由断言

Between 路由断言有两个参数——datetime1 和 datetime2。此断言匹配 datetime1 之后和 datetime2 之前发生的请求。datetime2 参数时间必须在 datetime1 之后。

application.yml 配置如下：

```
spring:
    cloud:
        gateway:
            routes:
            - id: between_route
```

```
            uri: http://example.org
            predicates:
            - Between=2017-01-20T17:42:47.789-07:00[America/Denver],
              2017-01-21T17:42:47.789-07:00[America/Denver]
```

4. Cookie 路由断言

Cookie 路由断言有两个参数，cookie 名称和正则表达式。此断言匹配具有给定名称且值与正则表达式匹配的 cookie。

application.yml 配置如下：

```
spring:
    cloud:
        gateway:
            routes:
            - id: cookie_route
              uri: http://example.org
              predicates:
              - Cookie=chocolate, ch.p
```

此路由匹配请求具有名为 chocolate 的 cookie，其值与 ch.p 正则表达式匹配。

5. Headers 断言工厂

Header 断言工厂采用两个参数，Headers 名称和正则表达式。此断言与具有给定名称且值与正则表达式匹配的 HTTP 头。

application.yml 配置如下：

```
spring:
    cloud:
        gateway:
            routes:
            - id: header_route
              uri: http://example.org
              predicates:
              - Header=X-Request-Id, \d+
```

如果请求具有名为 X-Request-Id 的 HTTP 头，则与该路由匹配，其值与 \d+ 正则表达式匹配（具有一个或多个数字的值）。

6. Host 路由断言

Host 路由断言采用一个参数：Host 值。

application.yml 配置如下：

```
spring:
    cloud:
        gateway:
            routes:
            - id: host_route
              uri: http://example.org
```

```
                predicates:
                - Host=**.somehost.org
```

如果请求的 Host 具有值 www.somehost.org 或 beta.somehost.org，则此路由将匹配。

7. Method 路由断言

Method 路由断言采用一个参数，即要匹配的 HTTP Method。

application.yml 配置如下：

```
spring:
    cloud:
        gateway:
            routes:
            - id: method_route
              uri: http://example.org
              predicates:
              - Method=GET
```

如果请求方法是 GET，则与此断言匹配。

8. Path 路由断言

Path 路由断言采用一个参数：Spring PathMatcher 模式。

application.yml 配置如下：

```
spring:
    cloud:
        gateway:
            routes:
            - id: host_route
              uri: http://example.org
              predicates:
              - Path=/foo/{segment}
```

如果请求路径是 /foo/1 或 /foo/bar，则匹配路由规则。

9. Query 路由断言

Query 路由断言有两个参数：一个必填的 param 和一个可选的 regexp。

application.yml 如下：

```
spring:
    cloud:
        gateway:
            routes:
            - id: query_route
              uri: http://example.org
              predicates:
              - Query=baz
```

如果请求包含 baz 查询参数，则此路由将匹配。

application.yml 的另一个示例：

```yaml
spring:
  cloud:
    gateway:
      routes:
      - id: query_route
        uri: http://example.org
        predicates:
        - Query=foo, ba.
```

如果请求包含 foo 查询参数且其值与 ba. 这个正则表达式匹配，则此路由将匹配，因此 bar 和 baz 将匹配。

10. RemoteAddr 路由断言

RemoteAddr 路由断言能够对 IP 进行过滤。例如在 192.168.0.1/16 的 IP 段内则匹配。
application.yml 配置如下：

```yaml
spring:
  cloud:
    gateway:
      routes:
      - id: remoteaddr_route
        uri: http://example.org
        predicates:
        - RemoteAddr=192.168.1.1/24
```

如果请求的远程地址是 192.168.1.10，则此路由将匹配。

这里有个特殊情况，如果 Spring Cloud Gateway 位于代理层（如 Nginx 等）后面，则可能与实际客户端 IP 地址不匹配。

我们可以通过自行实现 RemoteAddressResolver 来自定义解析远程地址的方式。

Spring Cloud Gateway 也附带一个非默认的解析器 XForwardedRemoteAddressResolver。它有两个静态构造函数方法。

1）XForwardedRemoteAddressResolver::trustAll，返回 RemoteAddressResolver，它始终采用 X-Forwarded-For 头中找到的第一个 IP 地址。但因为恶意客户端可以自行为解析器接收的 X-Forwarded-For 设置初始值，所以安全方面是存在隐患的。

2）XForwardedRemoteAddressResolver::maxTrustedIndex 采用与 Spring Cloud Gateway 前置运行的代理层数量相关的值表示。例如，如果只能通过 HAProxy 访问 Spring Cloud Gateway，则应使用值 1。如果在可访问 Spring Cloud Gateway 之前有两层代理层，则应使用值 2。

例如，给出 HTTP 头值为：

```
X-Forwarded-For: 0.0.0.1, 0.0.0.2, 0.0.0.3
```

则 XForwardedRemoteAddressResolver 的构造函数中如果传入不同的 maxTrustedIndex

值，将对应产出以下不同远程地址，如表 8-1 所示。

表 8-1

maxTrustedIndex	结果
[Integer.MIN_VALUE,0]	(invalid, IllegalArgumentException during initialization)
1	0.0.0.3
2	0.0.0.2
3	0.0.0.1
[4, Integer.MAX_VALUE]	0.0.0.1

同样可以使用 GatewayConfig.java 配置：

```
RemoteAddressResolver resolver = XForwardedRemoteAddressResolver
    .maxTrustedIndex(1);
.route("direct-route",
    r -> r.remoteAddr("10.1.1.1", "10.10.1.1/24")
        .uri("https://downstream1"))
.route("proxied-route",
    r -> r.remoteAddr(resolver,  "10.10.1.1", "10.10.1.1/24")
        .uri("https://downstream2"))
)
```

8.3　过滤器

路由过滤器允许以某种方式修改传入的 HTTP 请求或传出的 HTTP 响应。Spring Cloud Gateway 包含许多内置的 GatewayFilter 工厂，下面逐一学习每个过滤器的场景与使用方法。

1. AddRequestHeader 过滤器

AddRequestHeader 过滤器能够在请求头中新增信息，接收头名称和值两个参数。

我们在 application.yml 进行如下配置：

```yaml
spring:
    cloud:
        gateway:
            routes:
            - id: add_request_header_route
              uri: http://example.org
              filters:
              - AddRequestHeader=X-Request-Foo, Bar
```

这将为所有匹配请求的下游请求头添加 X-Request-Foo:Bar 键值对。

2. AddRequestParameter 过滤器

AddRequestParameter 过滤器，能够在请求参数中添加参数键值对，该过滤器接收名称和值两个参数。

application.yml 添加如下配置：

```yaml
spring:
  cloud:
    gateway:
      routes:
      - id: add_request_parameter_route
        uri: http://example.org
        filters:
        - AddRequestParameter=foo, bar
```

这会将 foo=bar 添加到所有匹配请求的查询参数字符串中。

3. AddResponseHeader 过滤器

AddResponseHeader 过滤器能够在响应头中添加键值对，该过滤器接收名称和值两个参数。我们在 application.yml 添加如下配置：

```yaml
spring:
  cloud:
    gateway:
      routes:
      - id: add_request_header_route
        uri: http://example.org
        filters:
        - AddResponseHeader=X-Response-Foo, Bar
```

这会将 X-Response-Foo:Bar 键值对添加到所有匹配请求的响应头中。

4. Hystrix 过滤器

Hystrix 是来自 Netflix 的熔断组件，我们在前文章节中已经学习过，该过滤器通过 circuit breaker pattern 实现。Hystrix 过滤器允许我们将断路器引入网关路由，保护服务免受雪崩故障的影响，并允许我们在下游故障时提供故障转移响应。

我们需要在 Maven 的 pom 文件中添加对 spring-cloud-starter-netflix-hystrix 的依赖。Hystrix 过滤器需要一个 name 参数，该参数是 HystrixCommand 的名称。

application.yml 文件配置如下：

```yaml
spring:
  cloud:
    gateway:
      routes:
      - id: hystrix_route
        uri: http://example.org
        filters:
        - Hystrix=myCommandName
```

这将使名为 myCommandName 的 HystrixCommand 包含在过滤器链表中。

5. PrefixPath 过滤器

PrefixPath 过滤器采用单个 prefix 参数，它能对请求路径进行前缀匹配修改。
application.yml 配置如下：

```yaml
spring:
  cloud:
    gateway:
      routes:
      - id: prefixpath_route
        uri: http://example.org
        filters:
        - PrefixPath=/mypath
```

该配置表示为所有匹配请求的路径添加 /mypath 前缀。因此，对 /hello 的请求将被发送到 /mypath/hello。

6. RedirectTo 过滤器

RedirectTo 过滤器接收 status 和 url 两个参数。表示将所有匹配的路径进行重定向操作。application.yml 配置如下：

```yaml
spring:
  cloud:
    gateway:
      routes:
      - id: prefixpath_route
        uri: http://example.org
        filters:
        - RedirectTo=302, http://acme.org
```

上述配置表示将发送 HTTP 状态 302 响应，以执行向 http://acme.org 的重定向操作。

7. RemoveNonProxyHeaders 过滤器

RemoveNonProxyHeaders 过滤器能够从转发的请求中删除代理层相关的 HTTP 头。该过滤器默认删除如下 HTTP 头参数。

- Connection
- Keep-Alive
- Proxy-Authenticate
- Proxy-Authorization
- TE
- Trailer
- Transfer-Encoding

如果要更改此删除列表，我们可以将 spring.cloud.gateway.filter.remove-non-proxy-headers.headers 属性设置为需要删除的 Headers 名称列表。

8. RemoveRequestHeader 过滤器

RemoveRequestHeader 过滤器接收一个 name 参数，它表示要删除的 HTTP 请求头的名称。application.yml 配置如下：

```yaml
spring:
  cloud:
    gateway:
      routes:
      - id: removerequestheader_route
        uri: http://example.org
        filters:
        - RemoveRequestHeader=X-Request-Foo
```

这个配置表明将在向下游发送之前删除 X-Request-Foo 标头。

9. RemoveResponseHeader 过滤器

RemoveResponseHeader 接收一个 name 参数，以删除响应报文中的 HTTP 头名称。application.yml 的配置如下：

```yaml
spring:
  cloud:
    gateway:
      routes:
      - id: removeresponseheader_route
        uri: http://example.org
        filters:
        - RemoveResponseHeader=X-Response-Foo
```

这个配置将在响应返回到客户端之前从响应中删除 X-Response-Foo 标头。

10. RewritePath 过滤器

RewritePath 过滤器接收路径 regexp 参数和 replacement 参数。该过滤器支持 Java 正则表达式来灵活地重写请求路径。application.yml 配置如下：

```yaml
spring:
  cloud:
    gateway:
      routes:
      - id: rewritepath_route
        uri: http://example.org
        predicates:
        - Path=/foo/**
        filters:
        - RewritePath=/foo/(?<segment>.*), /$\{segment}
```

对于 /foo/bar 的请求路径，这将在发出下游请求之前将路径设置为 /bar。

11. SaveSession 过滤器

SaveSession 过滤器支持在转发下游之前强制执行 WebSession::save 操作，确保在转发调用之前已保存会话状态。

application.yml 配置如下：

```
spring:
  cloud:
    gateway:
      routes:
      - id: save_session
        uri: http://example.org
        predicates:
        - Path=/foo/**
        filters:
        - SaveSession
```

12. SetPath 过滤器

SetPath 过滤器提供了一种允许模板化路径段来操作请求路径的简单方法。它使用了 Spring Framework 中的 uri 模板，允许有多个匹配的段。

application.yml 配置如下：

```
spring:
  cloud:
    gateway:
      routes:
      - id: setpath_route
        uri: http://example.org
        predicates:
        - Path=/foo/{segment}
        filters:
        - SetPath=/{segment}
```

/foo/bar 的请求路径将在发出下游请求之前将路径设置为 /bar。

13. SetResponseHeader 过滤器

SetResponseHeader GatewayFilter Factory 接受 name 和 value 参数，用来设置 HTTP 响应头。

application.yml 配置如下：

```
spring:
  cloud:
    gateway:
      routes:
      - id: setresponseheader_route
        uri: http://example.org
        filters:
        - SetResponseHeader=X-Response-Foo, Bar
```

14. SetStatus 过滤器

SetStatus 过滤器接收单个 status 参数，表示 HTTP 响应码。当然必须是有效的 Spring

HttpStatus，如整数值 404 或枚举 NOT_FOUND 的字符串表示形式。

application.yml 配置如下：

```yaml
spring:
  cloud:
    gateway:
      routes:
      - id: setstatusstring_route
        uri: http://example.org
        filters:
        - SetStatus=BAD_REQUEST
      - id: setstatusint_route
        uri: http://example.org
        filters:
        - SetStatus=401
```

该配置表示在任何一种情况下，响应的 HTTP 状态都将设置为 401。

15. StripPrefix 过滤器

StripPrefix 过滤器接收单个参数 parts，表示在向下游发送请求之前从请求路径中剔除的第几段的内容。application.yml 配置如下：

```yaml
spring:
  cloud:
    gateway:
      routes:
      - id: nameRoot
        uri: http://nameservice
        predicates:
        - Path=/name/**
        filters:
        - StripPrefix=2
```

当通过网关发出请求 /name/bar/foo 时，路径将会被替换为 /name/foo。

16. Retry 过滤器

Retry 过滤器接收 retries、statuses、methods 和 series 作为参数。

- retries：重试次数。
- statuses：应重试的 HTTP 状码，使用 org.springframework.http.HttpStatus 表示。
- methods：应该重试的 HTTP 方法，使用 org.springframework.http.HttpMethod 表示。
- series：要重试的一系列状态代码，使用 org.springframework.http.HttpStatus.Series 表示。

application.yml 配置如下：

```yaml
spring:
  cloud:
    gateway:
```

```yaml
routes:
- id: retry_test
  uri: http://localhost:8080/flakey
  predicates:
  - Host=*.retry.com
  filters:
  - name: Retry
    args:
      retries: 3
      statuses: BAD_GATEWAY
```

在生产使用的过程中，我们不可避免地需要考虑除了实现业务功能外的一些诸如性能安全等方面的考虑，那么接下来一起学习 Gateway 中提供了哪些相关的支持。

17. TLS / SSL

我们可以通过配置使 Gateway 支持兼容 HTTPS 请求。我们先在 application.yml 新增如下配置，表示开启 SSL 支持并配置上秘钥的基本信息：

```yaml
server:
  ssl:
    enabled: true
    key-alias: scg
    key-store-password: scg1234
    key-store: classpath:scg-keystore.p12
    key-store-type: PKCS12
```

如果路由到 HTTPS 下游，则可以配置为信任具有以下配置的所有下游证书，application.yml 配置如下：

```yaml
spring:
  cloud:
    gateway:
      httpclient:
        ssl:
          useInsecureTrustManager: true
```

当然信任所有下游证书在生产环境中可能存在安全隐患。那么我们可以使用以下配置来配置一组已知可信任的证书：

```yaml
spring:
  cloud:
    gateway:
      httpclient:
        ssl:
          trustedX509Certificates:
          - cert1.pem
          - cert2.pem
```

> **注意** 如果 Spring Cloud Gateway 未配置受信任证书，则使用默认信任库（可以使用系统属性 javax.net.ssl.trustStore 覆盖）。

当 Gateway 通过 HTTPS 进行通信时，客户端会启动 TLS 握手。这次握手会有很多超时。我们可以通过如下方式配置这些超时时间：

```
spring:
    cloud:
        gateway:
            httpclient:
                ssl:
                    handshake-timeout-millis: 10000
                    close-notify-flush-timeout-millis: 3000
                    close-notify-read-timeout-millis: 0
```

8.4 小结

根据本章节的学习，相信大家已经对 Gateway 有了一个全面的了解。尽管 Gateway 相比 Zuul 多了很多新特性，性能上可能比 Zuul 更有优势，但是目前 Gateway 脱离孵化器还不久，项目本身还不够成熟，并未接受过大面积使用的考验，所以建议大家在网关选型的时候还是需要充分考虑生产环境。

第 9 章 调用链追踪：Spring Cloud Sleuth

随着分布式服务架构的流行，特别是微服务等设计理念在系统中的应用，业务的调用链越来越复杂，如图 9-1 所示。

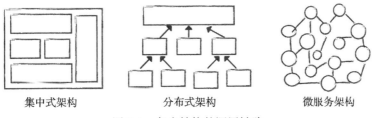

图 9-1 各个结构的调用链路

可以看到，随着服务的拆分，系统的模块变得越来越多，不同的模块可能由不同的团队维护。一个请求可能会涉及几十个服务的协同处理，牵扯到多个团队的业务系统，那么如何快速准确地定位到线上故障？

同时，缺乏一个自上而下全局的调用 id，如何有效地进行相关的数据分析工作？

一个典型的分布式系统请求调用过程如图 9-2 所示。

比较成熟的解决方案是通过调用链的方式，把一次请求调用过程完整串联起来，这样就实现了对请求调用路径的监控。

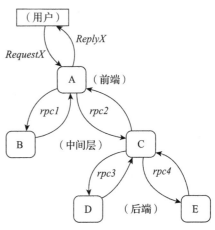

图 9-2 典型的分布式系统请求调用过程

一个调用链跟踪监控系统可以实现如下几点功能。

- **故障快速定位**：通过调用链跟踪，一次请求的逻辑轨迹可以完整清晰地展示出来。开发时可以在业务日志中添加调用链 ID，通过调用链结合业务日志快速定位错误信息。
- **各个调用环节的性能分析**：在调用链的各个环节分别添加调用时延，可以分析系统的性能瓶颈，进行针对性的优化。
- **各个调用环节的可用性，持久层依赖**：通过分析各个环节的平均时延、QPS 等信息，可以找到系统的薄弱环节，对一些模块做调整，如数据冗余等。
- **数据分析**：调用链是一条完整的业务日志，可以得到用户的行为路径，汇总分析应用在很多业务场景。

Spring Cloud Sleuth 就是 Spring Cloud 生态中实现调用链追踪的一个子项目，Spring Cloud Sleuth 可以结合 Zipkin，将信息发送到 Zipkin，利用 Zipkin 来存储信息，利用 Zipkin UI 展示数据，也可以只是简单地将数据记在日志中。

9.1 术语解释

Spring Cloud Sleuth 借用了 Dapper（Google 生产环境下的分布式跟踪系统）的术语。

- **Span**：最小工作单元。例如，发送一个 RPC 远程调用请求就是一个新的 Span，对 RPC 请求的响应也是一个新的 Span。Spring Cloud Sleuth 会为其生成一个 64 位的全局唯一的 ID。一个完整 Span 还包含描述、标注了时间的事件、标签、Span ID、进程 ID（通常是 IP 地址）。初始化的 Span 开始日志记录的时候被称作 Root Span。Span 的 ID 和日志 ID 的值是相等的。
- **Trace**：一系列的 Span 构成了一个树状结构，表示一次完整的调用链请求。
- **Annotation**：用来及时记录事件的存在，有些重要的注解用来定义请求的开始和结束。

一次完整的调用链请求分为如下几个注解。

- **cs - Client Request**：客户端发送了一个请求。这个注解描述一个 Span 的开始。
- **sr - Server Received**：服务端接收请求，并开始处理。从客户端请求发出到服务端接受收时间之间的时间差，表示网络延迟时间。
- **ss - Server Sent**：标识为请求处理完成（开始对客户端发送出响应）。从服务端接收到请求到这个时间点之间，表示服务端需要处理请求的时间。
- **cr - Client Received**：表示 Span 的结束，客户端已经从服务端成功接收请求。从这个时间点到服务端响应之间，表示客户端从服务端接收并响应需要的时间。

一次完整的调用链实例如图 9-3 所示，读者需要着重关注一下 TranceId、SpanId 和 Annotation 的变化。

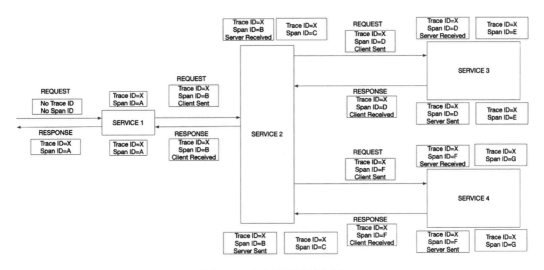

图 9-3　完整的调用链实例

9.2　Zipkin 简介

　　Zipkin 是一款开源的分布式实时数据追踪系统，基于 Google Dapper 的论文设计而来，由 Twitter 公司开发并贡献。其主要功能是聚集来自各个异构系统的实时监控数据，以追踪微服务架构下的系统延时问题。Spring Cloud Sleuth 打印日志并聚集至 Zipkin 中。Zipkin 主要由四部分构成：收集器、数据存储、查询以及 Web 界面。Zipkin 的收集器负责将各系统报告过来的追踪数据进行接收；而数据存储默认使用内存存储，也可以替换为 MySQL、Cassandra 等；查询服务用来向其他服务提供数据查询的能力，而 Web 服务是官方默认提供的一个图形用户界面。整体关系如图 9-4 所示。

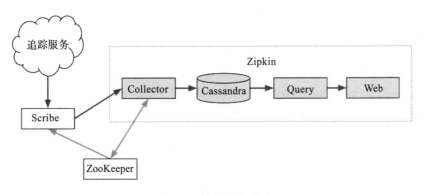

图 9-4　数据传递关系

　　在监控 Web 界面中，可以对监控的服务进行筛选，或者全量查看，如图 9-5 所示。

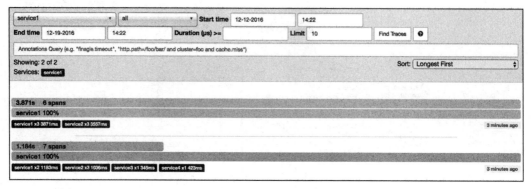

图 9-5　监控首页

一个完整的 Trace 在 Web 监控界面以如图 9-6 形式展示。

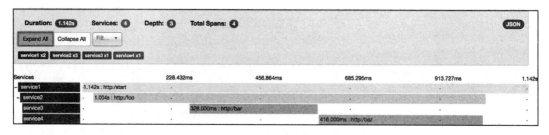

图 9-6　完整的 Trace 在 Web 监控界面

图 9-6 中的 Services 模块分别有 4 个 Service，对应的调用链表示如下。

1）service1 从 http:/start 路径接收到请求，表示为 2 个 Span，分别为 Server Received（SR）和 Server Sent（SS）annotations。

2）service1 调用 service2 的 http:/foo 路径。在 service1 中包含着 Client Sent（CS）和 Client Received 两个注解。在 service2 中包含着 Server Received 和 Server Sent 注解。

3）service2 调用 service3 的 http:/bar。在 service2 中包含着 Client Sent 和 Client Received annotations。在 service3 中包含着 Server Received 和 Server Sent 注解。

4）service2 调用 service4 的 http:/baz。在 service2 中包含着 Client Sent 和 Client Received annotations。在 service4 中包含着 Server Received 和 Server Sent 注解。

所以这个调用链一共包含了 7 个 Span：一个从 http://start 发起的 Span，两个 service1 调用 service2 产生的 Span，2 个从 service2 调用 service3 的 Span 和 2 个 service2 调用 service4 的 Span。

Trace 列表中红色信息是对错误信息的展示，点击 Trace 可以看到关于错误的详细信息，如图 9-7 所示。

第 9 章 调用链追踪：Spring Cloud Sleuth 93

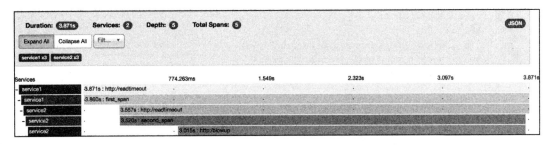

图 9-7　错误的详细信息

点击其中的 Span 可以看到关于错误信息的具体说明，以及这次请求的时间 HTTP 上下文等信息。

点击上方菜单栏中 Dependencies 可以看到服务调用关系的依赖图。图 9-8 为示例中的依赖图。

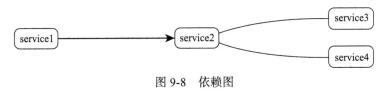

图 9-8　依赖图

9.3　使用 Zipkin

1. 启动 Zipkin

新建一个 Maven 项目，在 pom.xml 中 dependencyManagement 与其他示例相同，不再赘述，并在项目中添加如下依赖：

```
<dependency>
    <groupId>io.zipkin.java</groupId>
    <artifactId>zipkin-autoconfigure-ui</artifactId>
</dependency>
<dependency>
    <groupId>io.zipkin.java</groupId>
    <artifactId>zipkin-server</artifactId>
</dependency>
```

其中 zipkin-server 为 Zipkin 服务端的依赖，zipkin-autoconfigure-ui 为配置页面的依赖。接着在 application.yml 中指定端口：

```
server:
    port:8242
```

然后在主类中标注 @EnableZipkinServer，表示允许启动 Zipkin 服务端：

```java
@SpringBootApplication
@EnableZipkinServer
public class MysleuthserverApplication {
    public static void main(String[] args) {
        SpringApplication.run(MysleuthserverApplication.class, args);
    }
}
```

最后启动主类，并访问 http://localhost:8242/，就可以看到 Zipkin 监控页面了。

2. 在应用中添加 Sleuth

启动 Zipkin 服务端后，在 Eureka 示例和 Feign 示例中修改这两个应用的 pom.xml，增加对 Sleuth 的依赖。

```xml
<dependency>
    <groupId>org.springframework.cloud</groupId>
    <artifactId>spring-cloud-sleuth-zipkin</artifactId>
</dependency>
<dependency>
    <groupId>org.springframework.cloud</groupId>
    <artifactId>spring-cloud-starter-sleuth</artifactId>
</dependency>
```

其中 spring-cloud-starter-sleuth 会为应用日志生成并打印出 SpanId 和 TraceId。spring-cloud-sleuth-zipkin 提供了将日志以 HTTP 的形式传递到 Zipkin 服务端的能力。

然后分别在 provider 和 feignClient 对应的 application.yml 中添加 sleuth 和 zipkin 的配置：

```yaml
spring:
    zipkin:
        base-url: http://localhost:8242/
        enabled: true
    sleuth:
        sampler:
            percentage: 1.0
```

spring.zipkin.base-url 指定了 Zipkin 服务端的地址，而 spring.sleuth.sampler.percentage 则表示提取并传递日志的采样率，如 1.0 表示将全量 100% 的调用日志都传递到 Zipkin 服务器，在生产环境中可以将此比例调低，如 0.5。

接着我们尝试分别启动 provider 和 feignclient 应用并访问 feignclient 对应的 http://localhost:8081/sayHello?name=test，发起调用。然后访问 http://localhost:8242/，在 Zipkin 服务端的监控页面就可以看到 feignclient 对 provider 的调用信息了，与本章开头示例类似，不再贴图赘述。

在为应用添加完对 Sleuth 的依赖之后，仔细观察日志信息示例：

```
2016-02-02 15:30:57.902  INFO [bar,6bfd228dc00d216b,6bfd228dc00d216b,false]
    23030 --- [nio-8081-exec-3] ...
```

```
2016-02-02 15:30:58.372  ERROR [bar,6bfd228dc00d216b,6bfd228dc00d216b,false]
    23030 --- [nio-8081-exec-3] ...
2016-02-02 15:31:01.936   INFO [bar,46ab0d418373cbc9,46ab0d418373cbc9,false]
    23030 --- [nio-8081-exec-4] ...
```

可以看到日志信息前缀中多了如 [bar,46ab0d418373cbc9,46ab0d418373cbc9,false] 这一段日志信息。其分别表示 [appname,traceId,spanId,exportable]，第一个参数是服务名（对应 spring.application.name），第二个参数是 Trace ID，第三个参数是 Span ID，第四个标注 Sleuth 元信息是否导出。

3. 与 Spring Cloud Stream 整合

由于在日志传输到 Zipkin 服务端的过程中使用的是实时的 HTTP 请求，如果在访问量很高的生产环境中使用，不可避免地会带来很大的压力，如果能改成使用 Stream 消息机制的方式则可以大幅地提升性能。可以整合 Spring Cloud Stream 来使用消息机制传递用户的监控日志。

在 Zipkin 服务端的 pom.xml 中添加对 Stream 和消息中间件的依赖。接下来还是以 RabbitMQ 为例。

```xml
<dependency>
    <groupId>org.springframework.cloud</groupId>
    <artifactId>spring-cloud-sleuth-zipkin-stream</artifactId>
</dependency>
<dependency>
    <groupId>org.springframework.cloud</groupId>
    <artifactId>spring-cloud-sleuth-stream</artifactId>
</dependency>
<dependency>
    <groupId>org.springframework.cloud</groupId>
    <artifactId>spring-cloud-stream-binder-rabbit</artifactId>
</dependency>
```

在配置文件 application 中添加对 Stream 的配置信息和 RabbitMQ 的连接信息：

```yaml
spring:
  cloud:
    stream:
      binders:
        rabbit1:
          type: rabbit
          environment:
            spring:
              rabbitmq:
                host: localhost
                port: 5672
```

然后需要将 Zipkin 服务端主类的注解由 @EnableZipkinServer 替换为 @EnableZipkinStreamServer，@EnableZipkinStreamServer 注解包含了 @EnableZipkinServer，并增加了对

Stream 功能支持的能力，所以 Zipkin Server 仍然保留了从 HTTP 接收数据的能力。

接下来对 provider 和 feignclient 两个应用继续进行改造，以增加对 Stream 的支持。

我们分别在这两个项目的 pom.xml 中增加 Stream 和 binder 的依赖，去掉对 sleuth、Zipkin 的依赖。

```xml
<!--<dependency>-->
    <!--<groupId>org.springframework.cloud</groupId>-->
    <!--<artifactId>spring-cloud-sleuth-zipkin</artifactId>-->
<!--</dependency>-->
<dependency>
    <groupId>org.springframework.cloud</groupId>
    <artifactId>spring-cloud-sleuth-stream</artifactId>
</dependency>
<dependency>
    <groupId>org.springframework.cloud</groupId>
    <artifactId>spring-cloud-stream-binder-rabbit</artifactId>
</dependency>
```

然后与 Zipkin 服务端一样，在 application.yml 中增加 Stream 和 RabbitMQ 连接信息的配置。

读者可以自行启动并测试，可以看到实际运行功能与 HTTP 传递效果没有差别，只是传递方式变成了消息机制。

4. 与 MySQL 整合

Zipkin 对数据的存储默认是在内存中的。在应用的生产环境中，可能需要在数据库中持久化地存储这些数据。那我们同样可以对其整合。

在 Zipkin 服务端的 Maven 中增加 JDBC 和 MySQL 驱动的依赖：

```xml
<dependency>
    <groupId>org.springframework.boot</groupId>
    <artifactId>spring-boot-starter-jdbc</artifactId>
</dependency>
<dependency>
    <groupId>mysql</groupId>
    <artifactId>mysql-connector-java</artifactId>
    <optional>true</optional>
</dependency>
```

然后在配置文件中配置 MySQL 的链接信息：

```
application.yml
spring:
    datasource:
        schema: classpath:/mysql.sql
        url: jdbc:mysql://localhost/test
        username: root
```

```
password: root
initialize: true            #是否在启动时创建表结构
continueOnError: true
```

接下来可以看到应用启动时自动创建了表结构，并在有调用请求时自动进行了存储，表结构在图 9-9 中展示。其中 zipkin_annotations 完整记录了每一个注解的携带变量、对应的请求 IP 和端口以及请求的服务名等信息。zipkin_spans 记录了每一个 Span 的基本信息以及对应 Trace 的关联关系。zipkin_dependencies 记录了一些诸如日维度的调用次数等统计信息。

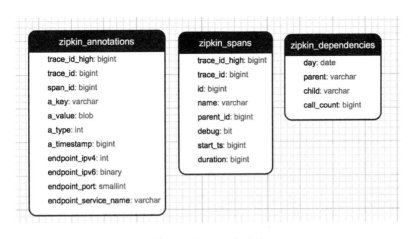

图 9-9　Zipkin 表结构

9.4　Span 进阶场景

9.4.1　自定义日志采样策略

分布式追踪日志数据量可能会非常大，因此在抽取日志的时候不可能全量抽取，而只需要根据一定的策略来采样即可很大程度地反映全貌。Spring Cloud Sleuth 的 Sampler 是所有采样策略的接口，可以实现这个接口来自定义策略控制抽样的算法。PercentageBasedSampler 是默认的抽样策略。可以通过 spring.sleuth.sampler.percentage 配置采样的比例。

我们可以通过 Spring 来定义一个 Bean，自定义 Sampler 配置所需要的抽样策略：

```
@Bean
public Sampler defaultSampler() {
    return new AlwaysSampler();
}
```

9.4.2 Span 的生命周期

可以通过 org.springframework.cloud.sleuth.Tracer 的接口对 Span 进行以下几种操作。
- start：开始对 Span 命名和记录开始时间戳。
- close：结束时记录结束时间戳并检查属性 exportable，然后汇总给 Zipkin，移除出当前的线程。
- continue：为 Span 新建实例并复制继续进行的 Span。
- detach：Span 没有 stop 或者 close，仅仅是移出当前的线程。
- create with explicit parent：在另外的一个线程重新创建一个 Span，并且明确它的父 Span。

 Spring 会为用户创建 Tracer。想要使用的话，只需要通过 Spring 注入它即可。

1. Span 的创建和结束
你可以用 Tracer 接口手动创建 Span。

```
//创建一个新的Span,如果当前线程中存在一个Span,则该Span会成为newSpan的父Span
Span newSpan = this.tracer.createSpan("calculateTax");
try {
    //添加一些自定义标签
    this.tracer.addTag("taxValue", taxValue);
    //也可以自定义一些监听事件
    newSpan.logEvent("taxCalculated");
} finally {
    //关闭当前Span,并将数据发往Zipkin
    this.tracer.close(newSpan);
}
```

在这个例子中可以看到如何创建一个 Span 实例。假设当前线程里面已经有一个 Span，它会成为这个 Span 的父节点。

2. 复用当前 Span
有时候你不想创建一个新的 Span，只希望继续使用原来的某个。可以通过如下方式实现。

```
Span continuedSpan = this.tracer.continueSpan(spanToContinue);
assertThat(continuedSpan).isEqualTo(spanToContinue);
//假设我们在线程Y中已经收到了线程X发来的'initialSpan'
Span continuedSpan = this.tracer.continueSpan(initialSpan);
try {
    this.tracer.addTag("taxValue", taxValue);
    continuedSpan.logEvent("taxCalculated");
} finally {
    //从当前线程中回收而非关闭continuedSpan
```

```
    this.tracer.detach(continuedSpan);
}
```

3. 父 Span 中创建新 Span

在指定的父 Span 下，创建一个新的 Span。通过 Tracer 接口下 startSpan 方法可以实现：

```
//假设工作在线程Y中，从线程X接收了'initialSpan'，则'initialSpan'将是'newSpan'的父节点
Span newSpan = this.tracer.createSpan("calculateCommission", initialSpan);
try {
    this.tracer.addTag("commissionValue", commissionValue);
    newSpan.logEvent("commissionCalculated");
} finally {
//在创建一个Span之后，一定要记得关闭Span
    this.tracer.close(newSpan);
}
```

9.4.3 重命名 Span

在一些具体的业务场景下实现链路追踪，我们会期望 Span 名称可以表达某些特定的业务意义，而不是由系统随机生成。可以根据如下步骤来一步步地实现自定义名称的 Span。

1. @SpanName 注解

可以用 @SpanName 注解指定 Span 的名称。

```
@SpanName("calculateTax")
class TaxCountingRunnable implements Runnable {
    @Override public void run() {
        //你的逻辑
    }
}
```

2. toString() 方法

很少的情况下需要单独为 Runnable 或者 Callable 创建类，一般情况是创建一个匿名实例。所以无法通过添加注解重写，没有 @SpanName 注解，这种场景下，我们可以通过重写 toString() 方法来实现。因此，通过如下代码，可以创建一个 calculateTax Span：

```
Runnable runnable = new TraceRunnable(tracer, spanNamer, new Runnable() {
    @Override public void run() {
    }
    @Override public String toString() {
        return "calculateTax";
    }
});
Future<?> future = executorService.submit(runnable);
future.get();
```

9.4.4 自定义 Span

可以使用 SpanInjector 和 SpanExtractor，来自定义 Span 的创建和发送，我们可以通过 HTTP 和 Spring 两种方式传递追踪信息。

1. Spring 方式

可以从 Message 创造 Span，并用 MessageBuilder 填充跟踪信息。你可以通过 @Primary 注解添加到自定义的对象上重写。

```
@Bean
public SpanExtractor<Message> messagingSpanExtractor() {
    ...
}
@Bean
public SpanInjector<MessageBuilder> messagingSpanInjector() {
    ...
}
```

2. HTTP 方式

我们可以从 HttpServletRequest 创建 Span，并用 HttpServletResponse 填充跟踪信息。

```
@Bean
public SpanExtractor<HttpServletRequest> httpServletRequestSpanExtractor() {
    ...
}
@Bean
public SpanInjector<HttpServletResponse> httpServletResponseSpanInjector() {
    ...
}
```

3. 示例

假设调用不是标准的 Zipkin 格式，只有兼容跟踪信息的 HTTP 头信息名称。那么，可以通过如下方式自定义 Span。参看如下 SpanExtractor 的例子：

```
static class CustomHttpServletRequestSpanExtractor
        implements SpanExtractor<HttpServletRequest> {
    @Override
    public Span joinTrace(HttpServletRequest carrier) {
        long traceId = Span.hexToId(carrier.getHeader("correlationId"));
        long spanId = Span.hexToId(carrier.getHeader("mySpanId"));
        Span.SpanBuilder builder = Span.builder().traceId(traceId).spanId(spanId);
        return builder.build();
    }
}
```

以下情况会创建 SpanInjector。

```
static class CustomHttpServletResponseSpanInjector
```

```
        implements SpanInjector<HttpServletResponse> {
    @Override
    public void inject(Span span, HttpServletResponse carrier) {
        carrier.addHeader("correlationId", Span.idToHex(span.getTraceId()));
        carrier.addHeader("mySpanId", Span.idToHex(span.getSpanId()));
    }
}
```

然后你还可以这样注册：

```
@Bean
@Primary
SpanExtractor<HttpServletRequest> customHttpServletRequestSpanExtractor() {
    return new CustomHttpServletRequestSpanExtractor();
}
@Bean
@Primary
SpanInjector<HttpServletResponse> customHttpServletResponseSpanInjector() {
    return new CustomHttpServletResponseSpanInjector();
}
```

4. 模拟发送调用日志

有的时候调用链路中的某个节点可能不支持 Sleuth，那么可以自定义创建一个 Span 标签，设置被调用方的相关信息。由以下例子可以看到，调用 Redis 服务的请求会被记录到 Span 里：

```
org.springframework.cloud.sleuth.Span newSpan = tracer.createSpan("redis");
try {
    newSpan.tag("redis.op", "get");
    newSpan.tag("lc", "redis");
    newSpan.logEvent(org.springframework.cloud.sleuth.Span.CLIENT_SEND);
    //例如调用Redis服务
    // return (SomeObj) redisTemplate.opsForHash().get("MYHASH", someObjKey);
} finally {
    newSpan.tag("peer.service", "redisService");
    newSpan.tag("peer.ipv4", "1.2.3.4");
    newSpan.tag("peer.port", "1234");
    newSpan.logEvent(org.springframework.cloud.sleuth.Span.CLIENT_RECV);
    tracer.close(newSpan);
}
```

9.5　其他场景与配置

1. 异步场景

如果在做业务处理时需要用到如线程、回调、异步等场景，那么使用 Sleuth 同样能记录下 Span。

Runnable 的例子：

```
Runnable runnable = new Runnable() {
    @Override
    public void run() {
    }
};
Runnable traceRunnable = new TraceRunnable(tracer, spanNamer, runnable,
"calculateTax");
//使用'Tracer'包装'Runnable'，Span名称将通过'@SpanName'注解或者'toString'方法获取
Runnable traceRunnableFromTracer = tracer.wrap(runnable);
```

Callable 的例子：

```
Callable<String> callable = new Callable<String>() {
    @Override
    public String call() throws Exception {
        return someLogic();
    }
    @Override
    public String toString() {
        return "spanNameFromToStringMethod";
    }
};
Callable<String> traceCallable = new TraceCallable<>(tracer, spanNamer, callable,
    "calculateTax");
Callable<String> traceCallableFromTracer = tracer.wrap(callable);
```

这种方式可以保证每次执行每个 Span 能被创建和关闭。

2. Hystrix 整合

（1）并发策略

通过注册自定义的 HystrixConcurrencyStrategy 的方式，把所有的 Callable 实例包装到 Sleuth-TraceCallable 中。并发策略开始或者继承 Span 取决于记录是在 Hystrix 命令之前还是之后。要禁用自定义的 Hystrix 并发策略，只要把 spring.sleuth.hystrix.strategy.enabled 设置为 false。

（2）命令设置

假定有以下 HystrixCommand 命令：

```
HystrixCommand<String> hystrixCommand = new HystrixCommand<String>(setter) {
    @Override
    protected String run() throws Exception {
        return someLogic();
    }
};
```

为了传递一些跟踪信息，则需要把某些操作的 TraceCommand 包装为 Sleuth 对应的 HystrixCommand 命令：

```
TraceCommand<String> traceCommand = new TraceCommand<String>(tracer, traceKeys,
setter) {
    @Override
    public String doRun() throws Exception {
        return someLogic();
    }
};
```

3. HTTP 过滤器

通过 TraceFilter 过滤的所有取样对内请求会创建 Span。Span 的名称为 "http:+ 请求的路径"。例如，请求的路径是 /foo/bar，名称则为 http:/foo/bar。可以通过 spring.sleuth.web.skipPattern 属性配置需要忽略的 URI。

4. Feign 整合

默认情况下，Spring Cloud Sleuth 通过 TraceFeignClientAutoConfiguration 提供对 Feign 的集成。

> **提示** 你可以通过设置 spring.sleuth.feign.enabled 为 false 禁用它，那么 Spring Cloud Sleuth 则不会加载任何自定义的 Feign 相关组件。

（1）@Async 和 @Scheduled 的支持

我们可以通过 @Async 注解方法，自动创建新的 Span。

Span 的名称会变成注解的方法名；Span 会有方法所在的类名的标签和方法名的标签。如果想忽略某些 @Scheduled 注解的方法自动创建 Span，可以通过设置 spring.sleuth.scheduled.skipPattern 为一个正则表达式匹配所有 @Scheduled 注解的方法名，阻止创建 Span。

（2）Executor、ExecutorService 和 ScheduledExecutorService 的支持

Spring 提供对 LazyTraceExecutor、TraceableExecutorService 和 TraceableScheduledExecutorService 的默认支持。会在新的任务提交、调用、计划触发时创建新的 Span。在用 CompletableFuture 时可以看到通过 TraceableExecutorService 传递跟踪消息的过程：

```
CompletableFuture<Long> completableFuture = CompletableFuture.supplyAsync(() -> {
    return 1_000_000L;
}, new TraceableExecutorService(executorService,
        tracer, traceKeys, spanNamer, "calculateTax"));
```

（3）Spring Integration 的支持

Spring Cloud Sleuth 也支持 Spring Integration 的集成。在 Integration 发布和订阅任务时，会自动创建 Span。要禁用对 Spring Integration 的支持，可以设置 pring.sleuth.integration.enabled:false。

默认情况下，所有的监控都会被捕获，可以使用 spring.sleuth.integration.patterns 配置

表达式进行过滤。

（4）Zuul 的支持

对 Zuul 的调用同样支持调用追踪。可以设置 spring.sleuth.zuul.enabled 为 false，禁用对 Zuul 的支持。

9.6 小结

在本章学习了 Sleuth 之后，读者是否已经迫不及待地思考如何将其应用于企业级环境当中了呢？业界整套日志采集分析展示的解决方案如 ELK、Flume 等。读者可以根据自身的实际场景像搭积木一样来组合选择适合自己日志采集、传输、分析、展现等各个环节的不同实现。

第 10 章 Chapter 10

加密管理：Vault

通常项目中敏感配置信息一般需要进行加密处理，比如数据库密码等，但 Spring Boot 内置不提供加密支持，不能加密配置文件信息。HashiCorp Vault 则可以解决这个问题，它提供集中管理密文和保护敏感数据的服务，而 Spring Cloud Vault 作为 HashiCorp Vault 的客户端，支持访问 HashiCorp Vault 内存储的数据，避免了在 Spring Boot 程序中存储敏感数据，为该场景提供了解决方案。它允许应用程序以透明的方式访问存储在 Vault 实例中的密文。比如 API 令牌、SSL 证书和口令。它还可以负责处理用户的访问控制，具有撤销令牌、审计功能，跟踪用户等功能。

10.1 初识 HashiCorp Vault

HashiCorp Vault 提供集中管理机密（Secret）和保护敏感数据的服务，可通过 UI、CLI 或 HTTP API 访问。HashiCorp Vault 使用 Go 语言编写。

1. 安装 HashiCorp Vault

根据大家的操作系统请自行在 https://www.vaultproject.io/downloads.html 下载 HashiCorp Vault 安装包，然后解压 zip 包，其中是一个可执行文件。以 Linux 系统为例：

```
$ unzip vault_1.0.2_linux_amd64.zip
$ sudo chown root:root vault
$ sudo chmod 755 vault
$ sudo mv vault /usr/local/bin/
$ vault --version
```

我们可以通过直接运行 vault 可查看支持的命令：

```
$ vault
Usage: vault <command> [args]
Common commands:
    read          Read data and retrieves secrets
    write         Write data, configuration, and secrets
    delete        Delete secrets and configuration
    list          List data or secrets
    login         Authenticate locally
    agent         Start a Vault agent
    server        Start a Vault server
    status        Print seal and HA status
    unwrap        Unwrap a wrapped secret
Other commands:
    audit         Interact with audit devices
    auth          Interact with auth methods
    kv            Interact with Vault's Key-Value storage
    lease         Interact with leases
    namespace     Interact with namespaces
    operator      Perform operator-specific tasks
    path-help     Retrieve API help for paths
    plugin        Interact with Vault plugins and catalog
    policy        Interact with policies
    secrets       Interact with secrets engines
    ssh           Initiate an SSH session
    token         Interact with tokens
```

运行 vault command -h 可查看命令支持的参数。其中 path-help 命令可以查看系统、Secret 引擎、认证方法等路径支持的配置，在实际应用中经常用到。比如：

```
$ vault path-help sys/
$ vault path-help database/
$ vault path-help database/roles
$ vault path-help aws/
$ vault path-help auth/token/
$ vault path-help auth/aws/
```

2. 自动完成

在 Linux 系统环境下，Vault 支持命令自动完成功能，安装后输入 vault [tab] 会显示命令提示，需执行以下命令安装：

```
$ vault -autocomplete-install
$ exec $SHELL
```

安装后将在 ~/.bashrc 内添加如下内容：

```
complete -C /usr/local/bin/vault vault
```

3. dev 模式启动 Vault

以 dev 模式启动不需任何配置，数据保存在内存中。

```
$ vault server -dev
```

控制台输出如下内容：

```
==> Vault server configuration:
Api Address: http://127.0.0.1:8200
Cgo: disabled
Cluster Address: https://127.0.0.1:8201
Listener 1: tcp (addr: "127.0.0.1:8200", cluster address: "127.0.0.1:8201", max_
    request_duration: "1m30s", max_request_size: "33554432", tls: "disabled")
Log Level: (not set)
Mlock: supported: true, enabled: false
Storage: inmem
Version: Vault v1.0.1
Version Sha: 08df121c8b9adcc2b8fd55fc8506c3f9714c7e61
WARNING! dev mode is enabled! In this mode, Vault runs entirely in-memory
    and starts unsealed with a single unseal key. The root token is already
    authenticated to the CLI, so you can immediately begin using Vault.
    You may need to set the following environment variable:

$ export VAULT_ADDR='http://127.0.0.1:8200'
The unseal key and root token are displayed below in case you want to seal/unseal
the Vault or re-authenticate.
Unseal Key: xSahEjtRQMMwbyBW6+rIzE2RRJ4d8X7BmAyPsSk63yE=
Root Token: s.5bnclu8POKx2WCxETB4u8RqF
Development mode should NOT be used in production installations!
```

其中，Unseal Key、Root Token 要保存下来。以 dev 模式启动 Vault 其状态是 unseal 的，不需要使用 Unseal Key 解封服务器。访问 Vault 需要使用 Root Token。建议将 Vault 服务器地址保存到环境变量 VAULT_ADDR 中，否则使用命令行访问 Vault 时需要指定 -address 参数。

4. 查看 Vault Server 状态

查看 Vault Server 状态示例如下：

```
$ vault status -address=http://127.0.0.1:8200
```

说明：-address 默认为 https://127.0.0.1:8200。

5. 登录 Vault

从浏览器登录 Vault，在地址栏输入 http://localhost:8200，如图 10-1 所示。

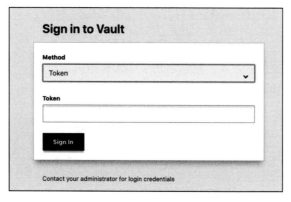

图 10-1　Vault 登录界面

在 Token 文本框内输入启动日志中生成的 Token：

```
Unseal Key: 37fPWUT1rc6SwxG2AeE/H9uE361vuric6EqalCTpOSo=
Root Token: s.BrsJKOmrShj0dYgFN2Pz4F56
```

进入 Vault 主界面，如图 10-2 所示。

图 10-2　Vault 主界面

从命令行登录 Vault：

```
$ vault login -method=token -address=http://127.0.0.1:8200
Token (will be hidden):
Success! You are now authenticated. The token information displayed below
is already stored in the token helper. You do NOT need to run "vault login"
again. Future Vault requests will automatically use this token.

Key                    Value
---                    -----
token                  s.1Pv48heTmZhXjm0bBd84Muef
token_accessor         3gfMlTXFPHX3ehMQzkJUrk3o
token_duration         ∞
token_renewable        false
token_policies         ["root"]
identity_policies      []
policies               ["root"]
```

6. 认证方法

Vault 支持多种登录认证方式，默认启用了 Token 方式。

从命令行查看启用的认证方法：

```
$ vault auth list
Path      Type     Accessor                  Description
----      ----     --------                  -----------
token/    token    auth_token_cd421269       token based credentials
```

7. Secret 引擎

Vault 支持多种 Secret 引擎，一些引擎只是提供存储和读取数据的支持，比如针对 key-value 数据结构的存取；一些引擎连接到其他服务并根据需要生成动态凭据，如 AWS 云服务、数据库；一些引擎提供加密服务（如 transit）、证书生成（如 pki）等。Vault 默认启用了 KV（Key-Value）引擎和 cubbyhole 引擎。

从命令行查看启用的 Secret 引擎：

```
$ vault secrets list
Path            Type         Accessor              Description
----            ----         --------              -----------
cubbyhole/      cubbyhole    cubbyhole_835f8a75    per-token private secret storage
identity/       identity     identity_0ba84c63     identity store
secret/         kv           kv_9558dfb7           key/value secret storage
sys/            system       system_5f7114e7       system endpoints used for control,
                                                   policy and debugging
```

我们在 Vault 引擎页面中 secret 分页下创建一个 secret 供后面测试使用，如图 10-3 所示。

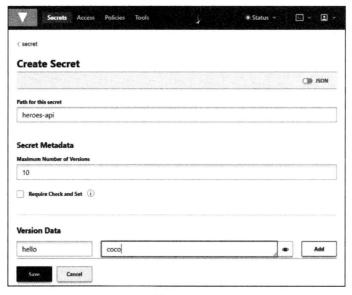

图 10-3

我们也可以使用命令行：

```
$ vault kv put secret/heroes-api hello=coco
```

查询 secret：

```
$ vault kv get secret/heroes-api
```

8. Token 和 Policy 管理

Root Token 具有最高权限，最佳实践不应存储 Root Token，仅在必要时使用 vault operator generate-root 命令生成，用毕撤销 Token。

1）撤销 Token。

```
$ vault token revoke -self
```

2）生成 Root Token。

① 初始化 Root Token，生成 one-time password（OTP）、Nonce。

```
$ vault operator generate-root -init
A One-Time-Password has been generated for you and is shown in the OTP field.
You will need this value to decode the resulting root token, so keep it safe.
Nonce          94e81220-dc59-16c5-1f08-180551cfa158
Started        true
Progress       0/3
Complete       false
OTP            kVpqIjLf7BZQgNUbEBAuQPikRk
OTP Length     26
```

② 生成 Root Token。

```
$ vault operator generate-root
Operation nonce: 94e81220-dc59-16c5-1f08-180551cfa158
Unseal Key (will be hidden):
Nonce          94e81220-dc59-16c5-1f08-180551cfa158
Started        true
Progress       1/3
Complete       false需要输入3次Unseal Key，成功后将输出Encoded Token：
Encoded Token  GHhHHBovfg9dEQAiASNhFiEFMT0DOjw+Gx4
```

③ 解码 Token。

```
$ vault operator generate-root -decode=GHhHHBovfg9dEQAiASNhFiEFMT0DOjw+Gx4
    -otp=kVpqIjLf7BZQgNUbEBAuQPikRk
```

3）创建 Token，设定有效时间，不指定 policy。

```
$ vault token create -ttl 10m
Key                     Value
---                     -----
token                   s.8DibgV8wlTJq3ygtcfK4ne2K
token_accessor          NuElYtSnxF51JXli3LC6XKHM
token_duration          10m
token_renewable         true
token_policies          ["root"]
identity_policies       []
policies                ["root"]
```

新 Token 为当前使用 Token 的子 Token，权限继承自当前使用的 Token。
过期后可 renew Token：

```
vault token renew s.8DibgV8wlTJq3ygtcfK4ne2K
```

4）创建 Token，指定 Policy。Policy 有如下几种权限。

```
# This section grants all access on "secret/*". Further restrictions can be
# applied to this broad policy, as shown below.
path "secret/*" {
    capabilities = ["create", "read", "update", "delete", "list"]
}

# Even though we allowed secret/*, this line explicitly denies
# secret/super-secret. This takes precedence.
path "secret/super-secret" {
    capabilities = ["deny"]
}
```

创建策略文件，仅允许读取路径 secret/heroes-api：

```
$ vi heroes-policy.hcl
```

heroes-policy.hcl 文件内容如下：

```
path "secret/heroes-api" {
    capabilities = ["read"]
}
```

上传策略：

```
$ vault policy write heroes heroes-policy.hcl
```

使用新策略创建 Token：

```
$ vault token create -policy=heroes
Key                  Value
---                  -----
token                s.1bJDHR7VuSaHfquqmoQREioA
token_accessor       FGufmiTSqWcEaiZAg9nuLkvx
token_duration       768h
token_renewable      true
token_policies       ["default" "heroes"]
identity_policies    []
policies             ["default" "heroes"]
```

默认 duration 为 768h，policy 为 "default" "heroes"。

使用新 Token 登录，查看 secret：

```
$ vault login s.1bJDHR7VuSaHfquqmoQREioA
$ vault kv get secret/heroes-api
```

10.2　整合 Spring Cloud Vault

为了在基于 Maven 的 Spring Boot 项目中引入 spring-cloud-vault 依赖，我们可以使用 starter 构件器，它将加载所有必需的依赖项。

除了主要的 starter 之外，我们还将包括 spring-vault-config-databases，它增加了对数据

库认证信息的支持：

```xml
<dependency>
    <groupId>org.springframework.cloud</groupId>
    <artifactId>spring-cloud-starter-vault-config</artifactId>
</dependency>
<dependency>
    <groupId>org.springframework.cloud</groupId>
    <artifactId>spring-cloud-vault-config-databases</artifactId>
</dependency>
```

1. 基础配置

为了正常工作，Spring Cloud Vault 需要一种方法来确定与 Vault 服务器通信的基本信息，以及如何对其进行身份验证。

我们在 bootstrap.yml 或者 bootstrap.properties 文件中配置必要的信息：

```yaml
# bootstrap.yml
spring:
    cloud:
        vault:
            uri: https://localhost:8200
            ssl:
                trust-store: classpath:/vault.jks
                trust-store-password: changeit
```

spring.cloud.vault.uri 属性指向 Vault 的 API 地址。由于测试环境使用带有自签名证书的 HTTPS，所以还需要提供包含其公钥的密钥存储库。

对于我们想在应用程序中使用的各种类型的加密方式，Spring Cloud Vault 也需要额外的配置。下面的部分描述了如何添加两个常见的加密类型支持：键值对和数据库凭证。

2. 使用一般的加密模块

下面介绍使用一般的加密模块来访问存储在 Vault 中的键值对的未定版本的加密数据。

假设已经在 classpath 中添加了 spring-cloud-starter-vault-config 的依赖，我们所要做的就是在应用程序的 bootstrap.yml 文件中添加如下属性：

```yaml
spring:
    cloud:
        vault:
            generic:
                enabled: true
                application-name: fakebank
```

在这种情况下，application-name 属性是可选的。如果不指定，Spring 将使用 spring.application.name 的值代替。

我们现在可以使用存储在 secret/fakebank 中的所有键值对作为任何其他环境的属性。下面的代码片段展示了如何读取存储在此路径下的 foo 键的值：

```
@Autowired Environment env;
public String getFoo() {
    return env.getProperty("foo");
}
```

3. 使用数据库加密模块

数据库加密模块允许 Spring 应用程序使用 Vault 创建动态生成的数据库凭证。Vault 将这些凭证注入标准的 spring.datasource.username 和 spring.datasource.password 属性，这样就可以通过常规的数据源进行选择。

为了在 Spring 应用程序中使用 Vault 生成的数据库凭证，我们必须将 spring-cloud-vault-config-databases 以及相应的 JDBC 驱动程序加入 Maven 的依赖中。

我们还需要在 bootstrap.yml 文件中添加一些属性，从而使其在应用程序中生效：

```
spring:
    cloud:
        vault:
            database:
                enabled: true
                    role: fakebank-accounts-rw
```

这里最重要的属性是 role 属性，它持有存储在 Vault 中的数据库角色名称。在程序启动期间，Spring 将连接 Vault，并请求来它创建具有相应角色的新凭证。

默认情况下，Vault 将在指定的时间内注销与这些凭证相关的特权。当然 Spring Cloud Vaulty 也支持自动更新与获得的凭证相关的租约。这样做，证书将在应用程序运行时保持长期有效。

现在，让我们参考如下示例。下面的代码片段是从 Spring 容器所管理的数据源 Bean 中获得一个新的数据库连接：

```
Connection c = datasource.getConnection();
```

我们发现，在代码中没有 Vault 使用的痕迹。所有的集成都发生在环境级别，所以代码可以很容易地像往常一样进行单元测试。

```
@Value("${hello}")
String name;
```

4. Vault 客户端 SSL 配置

在考虑到与 Vault 通信时的安全性问题，我们可以通过设置如下属性来配置 SSL，使得通信时使用 HTTPS 协议。参考如下配置：

```
spring.cloud.vault:
    ssl:
        trust-store: classpath:keystore.jks
        trust-store-password: changeit
```

- trust-store：设置证书存放的地址。
- trust-store-password：设置证书密码。

 注意　只有在 Apache HTTP 或 OkHttp 客户端被引入依赖时，才能应用配置 spring.cloud.vault.ssl.*。

10.3　认证模式

不同场景对安全性和身份验证有不同的要求。Vault 通过提供多种身份验证方法来支持这种需求。Spring Cloud Vault 支持令牌和 AppId 身份验证等多种认证方式，下面逐一来看一下。

1. 令牌认证

令牌身份验证是默认的身份验证方法。如果泄露了令牌，则黑客可以访问 Vault，从而可以访问目标客户端的密文了。

```
Example 103.1. bootstrap.yml
    spring.cloud.vault:
        authentication: TOKEN
        token: 00000000-0000-0000-0000-000000000000
```

其中 authentication 将此值设置为 TOKEN 选择令牌认证方法，而 token 设置要使用静态令牌，通过这种方式能够实现一个固定令牌的最简单的鉴权模式，当然这种方式由于安全隐患是不适用于生产环境的。

2. AppId 身份验证

Vault 支持的 AppId 身份验证，其中包含两个令牌。AppId 默认为静态配置的 spring.application.name。第二个令牌是 UserId，它是由应用程序中动态的部分表示，通常与运行时环境相关。Spring Cloud Vault 支持 IP 地址，MAC 地址和静态 UserId IP，以及 MAC 地址表示为十六进制编码的 SHA256 哈希。

下面示例是基于 IP 地址的 UserId，使用本地主机的 IP 地址。

```
spring.cloud.vault:
    authentication: APPID
    app-id:
        user-id: IP_ADDRESS
```

其中 authentication 将此值设置为 APPID 表示选择 AppId 身份验证方法，user-id 设置可选的值是 IP_ADDRESS，MAC_ADDRESS 或实现自定义的类名 AppIdUserIdMechanism。

从命令行生成 IP 地址 UserId 的相应命令是：

```
$ echo -n 192.168.99.1 | sha256sum
```

基于 MAC 地址的 UserId 从本地主机获取其网络设备。配置还允许指定 network-interface 以选择正确的网卡设备。network-interface 的值是可选的，可以是接口名称或接口索引（从 0 开始）。

```
spring.cloud.vault:
    authentication: APPID
    app-id:
        user-id: MAC_ADDRESS
        network-interface: eth0
```

其中，network-interface 设置网络接口以获取物理地址。

从命令行生成 IP 地址，UserId 的相应命令如下：

```
$ echo -n 0AFEDE1234AC | sha256sum
```

其中 MAC 地址为大写且没有冒号。

3. 自定义 UserId

UserId 生成是一种自定义生成策略的机制。我们可以将 spring.cloud.vault.app-id.user-id 设置为任何字符串，配置的值将用作静态 UserId。

更高级的方法允许我们将 spring.cloud.vault.app-id.user-id 设置为类名。此类必须实现 org.springframework.cloud.vault.AppIdUserIdMechanism 接口和 createUserId 方法。每次使用 AppId 进行身份验证以获取令牌时，Spring Cloud Vault 将通过调用 createUserId 来获取 UserId。我们在配置中加入如下属性：

```
spring.cloud.vault:
    authentication: APPID
    app-id:
        user-id: com.examlple.MyUserIdMechanism
```

并实现 AppIdUserIdMechanism 接口，来自定义生成策略：

```
public class MyUserIdMechanism implements AppIdUserIdMechanism {
    @Override
    public String createUserId() {
        String userId = ...
        return userId;
    }
}
```

4. TLS 证书身份验证

CERT 认证模块允许使用由 CA 签名或自签名的 SSL / TLS 客户端证书进行身份验证。

要启用 CERT 身份验证，我们需要配置包含客户端证书和私钥的 Java Keystore，以及将 spring.cloud.vault.authentication 设为 CERT，配置如下。

```
spring.cloud.vault:
    authentication: CERT
```

```
    ssl:
        key-store: classpath:keystore.jks
        key-store-password: changeit
        cert-auth-path: cert
```

5. Kubernetes 身份验证

Kubernetes 身份验证机制允许使用 Kubernetes 服务账户令牌对 Vault 进行身份验证。身份验证基于角色，角色绑定到服务账户名称和命名空间。

包含 pod 的服务账户的 JWT 令牌的文件将自动挂载在 /var/run/secrets/kubernetes.io/serviceaccount/token，使用配置如下：

```
spring.cloud.vault:
    authentication: KUBERNETES
    kubernetes:
        role: my-dev-role
        service-account-token-file: /var/run/secrets/kubernetes.io/serviceaccount/token
```

其中 role 表示设置的角色，service-account-token-file 表示设置包含 Kubernetes 服务账户令牌的文件的位置，默认为 /var/run/secrets/kubernetes.io/serviceaccount/token。

10.4　三方组件支持

Vault 官方也默认提供了针对许多开发过程中会使用过的三方组件的支持，无须开发人员自行实现，即可开箱机用。

Spring Cloud Vault 默认支持如下三方组件：

- 支持 JDBC 协议的数据库
- Apache Cassandra
- MongoDB
- MySQL
- PostgreSQL
- RabbitMQ
- AWS

关于其他三方组件的支持，由于配置极其类似，下文列举两个示例，其他不再赘述，请读者自行查阅。

1. Consul

Spring Cloud Vault 可以获取 HashiCorp Consul 的凭据。Consul 集成需要 spring-cloud-vault-config-consul 依赖项：

```
<dependencies>
<dependency>
```

```
        <groupId>org.springframework.cloud</groupId>
        <artifactId>spring-cloud-vault-config-consul</artifactId>
        <version>Finchley.SR2</version>
    </dependency>
</dependencies>
```

可以通过设置 spring.cloud.vault.consul.enabled=true（默认 false）并使用 spring.cloud.vault.consul.role=…提供角色名称来启用集成。

获取的令牌存储在 spring.cloud.consul.token 中，因此使用 Spring Cloud Consul 可以获取生成的凭据而无须进一步配置。读者也可以通过设置 spring.cloud.vault.consul.token-property 来配置属性名称。我们参考如下配置：

```
spring.cloud.vault:
    consul:
        enabled: true
        role: readonly
        backend: consul
        token-property: spring.cloud.consul.token
```

其中：
- enabled，将此值设置为 true，可启用 Consul 后端配置；
- role，设置 Consul 角色定义的角色名称；
- backend，设置要使用的 Consul 挂载路径；
- token-property，设置存储 Consul ACL 令牌的属性名称。

2. 数据库支持

Vault 支持多个数据库加密模块，以根据配置的角色动态生成数据库凭据。这意味着需要访问数据库的服务不再需要配置凭据，业务服务可以从 Vault 请求得到凭证，并使用 Vault 的租赁机制更轻松地刷新密钥。

我们如果需要引入 Spring Cloud Vault 对数据库的支持则需要在 Maven 的 pom.xml 中新增如下依赖：

```
<dependencies>
    <dependency>
        <groupId>org.springframework.cloud</groupId>
        <artifactId>spring-cloud-vault-config-databases</artifactId>
        <version>Finchley.SR2</version>
    </dependency>
</dependencies>
```

Spring Cloud Vault 可以通过设置 spring.cloud.vault.database.enabled=true（默认 false）并使用 spring.cloud.vault.database.role=…提供角色名称来启用集成。

spring.cloud.vault.database 可以支持 JDBC 协议的数据库，用户名与密码存储在 spring.datasource.username 和 spring.datasource.password 中，因此使用 Spring Boot 为 DataSource

获取生成的凭据，通过设置 spring.cloud.vault.database.username-property 和 spring.cloud.vault.database.password-property 来修改取值属性的名称。我们参考如下配置：

```
spring.cloud.vault:
    database:
        enabled: true
        role: readonly
        backend: database
        username-property: spring.datasource.username
        password-property: spring.datasource.username
```

- enabled，将此值设置为 true，表示启用数据库支持；
- role，设置数据库角色定义的角色名称；
- backend，设置要使用的数据库类型；
- username-property，设置存储数据库用户名的属性名称；
- password-property，设置存储数据库密码的属性名称。

10.5 小结

在学习完本章之后，大家应该能够熟练使用 Spring Cloud Vault 对敏感数据的保护惯例，那么接下来一起学习 Spring Cloud 还为我们提供了哪些实用的模块吧！

第 11 章 Chapter 11

公共子项目

11.1 命令行工具：Spring Boot CLI

Spring Boot CLI 是 Spring Boot 提供的一个命令行工具，它可以提供快速运行 Spring Boot 程序的功能，并支持插件功能，当我们为 Spring Boot CLI 安装上 Spring Cloud 插件时，它提供的命令行操作也可以用于 Spring Cloud。我们可以支持 Spring Cloud Config 客户端对配置信息的加解密处理，甚至可以直接通过 CLI 在命令行上运行诸如 Eureka、Zipkin、Config Server 等服务。它为我们日常开发中调试搭建环境测试等场景提供了极大的便利。

11.1.1 安装 Spring Boot CLI

步骤 1 安装 SDKMAN。

SDKMAN 是一个用来管理开发环境中各种不同依赖版本的工具。

1）在命令行运行：

```
curl -s "https://get.sdkman.io" | bash
```

2）等待下载执行完毕之后运行：

```
source "$HOME/.sdkman/bin/sdkman-init.sh"
```

3）测试安装结果：

```
sdk version
```

看到正常显示 SDK 版本号，则表示安装成功。

步骤 2 安装 Spring Boot CLI。

使用 SDKMAN 安装并管理 Spring Boot CLI 版本：

```
$ sdk install springboot
$ spring --version
Spring Boot v1.5.2.RELEASE
```

1）新建 MyController.java：

```
@Controller
public class MyController {
    @RequestMapping("/say")
    @ResponseBody
    public String  print(String name ){
        return "hello "+name;
    }
}
```

2）在文件当前目录下执行以下代码：

```
spring run MyController.java
```

可以看到控制台打印出一个常规 Spring Boot 的启动日志，读者可以自行尝试访问 localhost:8080/say?name=test 进行检测。如果需要依赖其他功能，也只需加上相应的注解，如 @EnableConfigServer、@EnableOAuth2Sso 或者 @EnableEurekaClient。

步骤 3　安装 Spring Boot Cloud 插件。

由于最新版本的 Spring Boot CLI v1.5.2.RELEASE 与 Spring Cloud CLI 插件的最新版本 1.2.3.RELEASE 有兼容问题，我们将使用 SDKMAN 管理更换 Spring Boot CLI 版本。

```
$ sdk install springboot 1.4.4.RELEASE
$ sdk use springboot 1.4.4.RELEASE
```

接着安装 Spring Cloud CLI 插件：

```
$ spring install org.springframework.cloud:spring-cloud-cli:1.4.0.RELEASE
```

11.1.2　使用 Spring Cloud CLI

在命令行执行 CLI 可以快速运行一个通用服务，如 Eureka、Config Server 等。可以通过命令罗列出支持的服务列表，也可以直接执行 Spring Cloud，这个命令会自动运行一系列默认服务。也可以通过命令指定需要启动的服务，例如：

```
$ spring cloud eureka configserver h2 kafka zipkin
```

我们可以看一下 help 命令的提示：

```
~ spring help cloud
spring cloud - Start Spring Cloud services, like Eureka, Config Server, etc.
Option            Description
------            -----------
-d, --debug       Debug logging for the deployer
-l, --list        List the deployables (don't launch
```

```
            anything)
    -v, --version   Show the version (don't launch
examples:
    Launch Eureka:
        $ spring cloud eureka
    Launch Config Server and Eureka:
        $ spring cloud configserver eureka
    List deployable apps:
        $ spring cloud --list
    Show version:
        $ spring cloud --version
```

❑ spring cloud –list：查看支持的服务；

❑ spring cloud –version：查看版本；

❑ spring cloud configserver eureka：启动 Config Server 和 Eureka。

下面简单罗列了一些支持的服务，参见表 11-1。

表 11-1 CLI 支持的所有服务及其对应默认值与解释

服务	名称	地址	描述
eureka	EurekaServer	http://localhost:8761	服务发现。默认情况下所有的服务都会注册
configserver	Config Server	http://localhost:8888	配置服务。使用"native"Profile，并且从本地"./launcher"目录加载服务配置
h2	H2 Database	http://localhost:9095 (console), jdbc:h2:tcp://localhost:9096/{data}	关系型数据库。连接时使用 {data} 作为 path，例如 ./target/test。可以在连接上添加；MODE=MYSQL 或；MODE=POSTGRESQL 进行自定义配置
kafka	Kafka Broker	http://localhost:9091 (actuator endpoints), localhost:9092	
hystrixdashboard	HystrixDashboard	http://localhost:7979	断路器。默认统计接口 /hystrix.stream
dataflow	DataflowServer	http://localhost:9393	Spring Cloud Dataflow 服务。带有一个 UI 界面：/admin-ui
zipkin	ZipkinServer	http://localhost:9411	Zipkin 服务

每一个应用都可以使用同一个名字的本地 YAML 文件进行配置（在当前工作目录，或一个名为 config 的子目录，或 ~/.spring-cloud）。例如：可以在 configserver.yml 文件中配置一个 Git 仓库，配置项如下：

```
spring:
    profiles:
        active: git
    cloud:
        config:
            server:
```

```
            git:
                uri: file://${user.home}/dev/demo/config-repo
```

11.1.3 加解密

Spring Cloud CLI 有一个"encrypt"和一个"decrypt"命令。这两个命令都接收参数"--key"，用于指定密钥。例如：

```
$ spring encrypt mysecret --key foo
682bc583f4641835fa2db009355293665d2647dade3375c0ee201de2a49f7bda
$ spring decrypt --key foo 682bc583f4641835fa2db009355293665d2647dade3375c0ee201
    de2a49f7bda
mysecret
```

如果需要使用密钥文件（如 RSA 公钥），可以用"@"来指定一个文件。例如：

```
$ spring encrypt mysecret --key @${HOME}/.ssh/id_rsa.pub
AQAjPgt3eFZQXwt8tsHAVv/QHiY5sI2dRcR+...
```

11.2 注册中心：Spring Cloud ZooKeeper

ZooKeeper 是 Apache Hadoop 的一个子项目，它主要用来解决分布式应用中经常遇到的一些数据管理问题，如统一命名服务、状态同步服务、集群管理、分布式应用配置项的管理等。ZooKeeper 的存储结构与传统的文件系统类似，通过一个树结构存储节点的形式来存储数据。

Spring Cloud ZooKeeper 则是封装了 ZooKeeper 来实现诸如注册中心、配置管理等功能，可以替换 Eureka、Spring Cloud Config 中的配置管理等功能。但是，路由网关（Zuul）、客户端负载均衡（Ribbon）、断路器（Hystrix）等还是继续由 Spring Cloud Netflix 提供。

11.2.1 安装 ZooKeeper

我们依旧使用 Docker 来启动 ZooKeeper，ZooKeeper 的集群支持和高可用等方面不是本书的侧重点，在此不再赘述。

```
docker run -p 2181:2181 -d zookeeper
```

启动后 ZooKeeper 对外暴露的连接端口为 2181。

11.2.2 基于 ZooKeeper 服务发现

Spring Cloud ZooKeeper 通过 Curator（ZooKeeper 的一个 Java 客户端类库）实现了服务注册和发现服务，实现了与 Eureka 等同的功能。

1. 服务提供方

我们对 Eureka 的服务提供方项目进行改造，在 Maven 的 pom.xml 中增加对 ZooKeeper

提供服务发现功能的类库的依赖，以及 Curator 关于服务发现功能的依赖。记得同时也要移除 Eureka 的相关依赖。

```xml
<dependency>
    <groupId>org.springframework.cloud</groupId>
    <artifactId>spring-cloud-zookeeper-discovery</artifactId>
</dependency>
<dependency>
    <groupId>org.apache.curator</groupId>
    <artifactId>curator-x-discovery</artifactId>
</dependency>
```

EurekaProviderApplication.java 主类本身代码结构不变。注意，我们对于开启自动服务发现的注解是要使用 @EnableDiscoveryClient，而不是 @EnableEurekaClient。这里就体现出我们使用抽象层次更高的注解而不是具体实现的注解的好处了。

在配置文件中增加关于 ZooKeeper 连接信息的配置。

application.yml 配置文件内容如下：

```yaml
spring:
    cloud:
        zookeeper:
            connect-string: localhost:2181
```

我们在启动以 ZooKeeper 为注册中心的服务提供者后，通过 ZooKeeper 的数据监控工具可以看到新增了如下节点，如图 11-1 所示。

图 11-1　ZooKeeper 中的注册数据

节点内容如下：

```
{
    "name":"myprovider",
    "id":"92eed180-8f8b-4068-a437-1eeed5478f76",
    "address":"192.168.14.27",
    "port":8085,
    "sslPort":null,
    "payload":{??
        "@class":"org.springframework.cloud.zookeeper.discovery.ZooKeeperInstance",
        "id":"myprovider:8085",
        "name":"myprovider",
        "metadata":{??
            "instance_status":"UP"
        }
```

```
        },
        "registrationTimeUTC":1494218112990,
        "serviceType":"DYNAMIC",
    }
}
```

可以看到,节点信息中记载了服务提供方的 IP、服务名称、节点状态等相关信息。

2. 服务消费方

同样,无须对代码进行改造。无论消费方使用的是 DiscoveryClient、Feign 还是 Ribbon,仅需跟服务提供方的改造一样,在 Maven 中添加 ZooKeeper 的依赖,在配置文件中增加 ZooKeeper 的连接信息即可。具体不再赘述,参考上文。

在启动服务消费方后,监控 ZooKeeper 数据同样也可以看到消费方的相关节点。同时,通过 DiscoveryClient、Feign 和 Ribbon 调用 /sayHello 均可以正常调用。

11.2.3 相关配置

通过下面这个例子来说明。定义配置文件 application.yml:

```yaml
spring.application.name: yourServiceName
spring.cloud.zookeeper:
    dependencies:
        newsletter:
            path: /path/where/newsletter/has/registered/in/zookeeper
            loadBalancerType: ROUND_ROBIN
            contentTypeTemplate: application/vnd.newsletter.$version+json
            version: v1
            headers:
                header1:
                    - value1
                header2:
                    - value2
            required: false
            stubs: org.springframework:foo:stubs
        mailing:
            path: /path/where/mailing/has/registered/in/zookeeper
            loadBalancerType: ROUND_ROBIN
            contentTypeTemplate: application/vnd.mailing.$version+json
            version: v1
            required: true
```

配置的根节点是 spring.cloud.zookeeper.dependencies。然后,一步一步地分析上面这个配置。

1. 服务 ID

从根节点往下的每一个配置都作为 Ribbon 的一个别名。别名的名称就是已经实例化的 DiscoveryClient、Feign 或者 RestTemplate 的 serviceId。

在上面的配置中，有两个别名：newsletter 和 mailing。例如，newsletter 使用 Feign 实现：

```
@FeignClient("newsletter")
public interface NewsletterService {
    @RequestMapping(method = RequestMethod.GET, value = "/newsletter")
    String getNewsletters();
}
```

2. Path
配置中的 path 属性表示注册在 ZooKeeper 中的位置。

3. Load balancer type
配置中的 loadBalancerType 属性表示调用这个节点时所采用的负载均衡策略。可以在下列值中选择一个。

- STICKY：固定（一直请求这个节点）；
- RANDOM：随机；
- ROUND_ROBIN：循环调用（排队）。

4. Content-Type 和 version
当应用 API 通过 Content-Type 头来传递版本信息时，如果每次请求都手工配置肯定不好，而且接口升级时还要重新编码制定版本信息也是极其低效的。所以，提供 contentTypeTemplate 并配有 $version 占位符，这样再配合 version 属性，就可以灵活地通过配置信息来指定版本信息了。例如，有类似 contentTypeTemplate:application/vnd.newsletter.$version+json 以及 version：v1 的信息，那么每一个请求的 Content-Type 就会自动设置为 application/vnd.newsletter.v1+json。

5. Headers 信息配置
配置中的 Headers 属性是一个 map 结构。有的时候，需要在每一次请求中设置一些头部信息。而这些都可以通过 Headers 属性在配置文件中进行配置，而不需要代码。例如：

```
headers:
    Accept:
        - text/html
        - application/xhtml+xml
    Cache-Control:
        - no-cache
```

这样就会在每一个 HTTP 请求的头部加入 Accept 和 Cache-Control。

6. Obligatory dependencies
这个配置可以指定在应用启动时，设置某些服务是不是必须要存活的。

如果设置为 true，则应用启动时不能发现服务会直接抛出异常，并且 Spring Context 启动失败。

7. Stubs 本地存根

可以配置一个用冒号分割的信息以指定本地存根（用于本地逻辑、降级、默认行为等）。可以通俗理解为指定一个 Backup，当服务降级、调用失败时，返回该存根依赖的 Mock 结果。

例如：

```
stubs: org.springframework:foo:stubs
```

其中具体含义如下。
- groupId：org.springframework；
- artifactId：foo；
- classifier：stubs? 默认值。

8. 其他配置

这部分主要介绍一些用以控制 ZooKeeper 特性的常用配置。

- spring.cloud.zookeeper.dependency.ribbon.enabled（默认开启）：可以针对 Ribbon 全局或者对某一个服务是否启用。
- spring.cloud.zookeeper.dependency.ribbon.loadbalancer（默认开启）：结合上面属性一起用的，可以针对运行时的负载均衡策略（loadBalancerType）。必要时，还可以自定义 LoadBalancerClient、ILoadBalancer 来自定义负载均衡策略。
- spring.cloud.zookeeper.dependency.headers.enabled（默认开启）：控制是否自动在请求头部加入信息。可以对诸如 RibbonClient 的每一次请求起作用。
- spring.cloud.zookeeper.dependency.resttemplate.enabled（默认开启）：类似上一个配置，控制着请求头信息。让带有 @LoadBalanced 注解的 RestTemplate 类，能够通过头信息携带的参数进行相应的负载策略。

11.2.4 节点监听

可以对 ZooKeeper 的节点注册监听器。当相关节点状态发生变化时，可以执行一些自定义逻辑。自定义监听器需要实现 org.springframework.cloud.zookeeper.discovery.watcher. DependencyWatcherListener 接口，并注册 Bean 到 Spring 容器中。这个接口主要是一个方法：

```
void stateChanged(String dependencyName, DependencyState newState);
```

如果需要对一个特定的节点进行监听，可以通过 dependencyName 进行鉴别。newState 则提供节点的 CONNECTED 或者 DISCONNECTED 的状态变化。

基于上面的依赖监听机制，Presence Checker 提供了应用启动时检查某些节点状态的能力。

- 如果节点配置成 required，而 ZooKeeper 中并没有注册信息，则抛出异常，应用退出；

❑ 如果节点配置为非 required，则 org.springframework.cloud.zookeeper.discovery. watcher.presence.LogMissingDependencyChecker 会记录一个 WARN 级别的日志，应用正常启动。

我们通过自定义 org.springframework.cloud.zookeeper.discovery.watcher.presence.DependencyPresenceOnStartupVerifier 以及 org.springframework.cloud.zookeeper.discovery.watcher.presence.DefaultDependencyPresenceOnStartupVerifier 两个类来实现对上述逻辑进行自定义。

11.3 注册中心：Spring Cloud Consul

Consul 是 HashiCorp 公司推出的开源工具，用于实现分布式系统的服务发现与配置。与其他分布式服务注册与发现的方案相比，Consul 的方案更"一站式"，内置了服务注册与发现框架、分布一致性协议实现、健康检查、Key/Value 存储、多数据中心方案，不再需要依赖其他工具（如 ZooKeeper 等）。

Spring Cloud Consul 是对 Consul 的封装，整合进 Spring Cloud 体系实现注册中心、分布式配置等功能。

这个项目在 Spring Boot 基础上，通过自动化配置，提供对 Consul 各组件的整合，并绑定到 spring 运行环境及各模块中。在分布式应用系统中，可以很方便地使用注解去启用配置强大的组件，实现分布式通用模式，如服务发现模式（Eureka）、熔断模式（Hystrix）、路由网关（Zuul）、客户端负载均衡模式（Ribbon）等。

11.3.1 安装 Consul

通过 Docker 运行 Consul 镜像：

```
$ docker run -p 8400:8400 -p 8500:8500 -p 8600:53/udp -h node1 progrium/consul
    -server -bootstrap -ui-dir /ui
```

上述代码暴露了 8400（RPC）、8500（HTTP）和 8600（DNS）三个端口。我们通过访问 http://localhost:8500/ui/ 可以看到其自带的监控平台。

11.3.2 基于 Consul 注册服务

依旧对上文 Eureka 的服务提供者和消费者案例进行改造，在 Maven 的 pom 中移除 Eureka 的依赖，并新增 Consul 的 Maven 依赖：

```
<dependency>
    <groupId>org.springframework.cloud</groupId>
    <artifactId>spring-cloud-consul-discovery</artifactId>
</dependency>
```

在配置文件 application.yml 中新增 Consul 的连接信息：

```yaml
spring:
  cloud:
    consul:
      host: localhost
port: 8500
```

在分别启动服务提供者和服务消费者案例之后，可以发现调用 Eureka 案例中已经实现过的 /sayHello 接口依旧能够正常调用并返回结果。

11.4 小结

本章介绍了一个客户端工具以及两个注册中心的实现。我们可以在具体的应用场景中根据不同的需要选择不同的注册中心，以及很方便地使用 Spring 为我们提供的 CLI 脚手架客户端。至此，本篇的内容就介绍完了，接下来将进入第二篇，介绍关于消息处理等方面的组件。

第二篇 *Part 2*

任务与消息篇

上一篇全面介绍了 Spring Cloud 生态中偏重于服务调用、服务治理、服务监控等方面的各个组件,接下来将进入一个新篇。这一篇将介绍 Spring Cloud 生态中有关消息总线、任务广播、批处理等方面的各个组件,同样非常实用。

第 12 章

消息驱动：Spring Cloud Stream

在企业级应用中处理非同步场景、消息通知、应用间解耦等场景经常会使用到消息中间件。我们可能已经使用过或者听过一些市面上常用的消息中间件，如 ActiveMQ、RabbitMQ、MetaQ、kafka、Redis 中的消息队列功能等。

Spring Cloud Stream 是一个构建消息驱动微服务的框架，基于 Spring Integration（https://projects.spring.io/spring-integration/）并利用 Spring Boot 提供了自动配置，支持发布 – 订阅模型、消费者分组、数据分片等功能，提供极为方便的消息中间件使用体验。

目前支持的中间件有以下这些。

- Apache Kafka（https://github.com/spring-cloud/spring-cloud-stream-binder-kafka）
- RabbitMQ（https://github.com/spring-cloud/spring-cloud-stream-binder-rabbit）
- JMS（https://github.com/spring-cloud/spring-cloud-stream-binder-jms）
- Redis（https://github.com/spring-cloud/spring-cloud-stream-binder-redis）
- Gemfire（https://github.com/spring-cloud/spring-cloud-stream-binder-gemfire）
- IBM-MQ（https://github.com/spring-cloud/spring-cloud-stream-binder-ibm-mq）
- Google-PubSub（https://github.com/spring-cloud/spring-cloud-stream-binder-google-pubsub）
- Solace（https://github.com/spring-cloud/spring-cloud-stream-binder-solace）

12.1　Stream 应用模型

Spring Cloud Stream 应用由第三方的中间件组成。应用间的通信通过输入通道（input channel）和输出通道（output channel）完成。这些通道是由 Spring Cloud Stream 注入的。而通道与外部的代理（可以理解为上文所说的数据中心）的连接又是通过 Binder 实现的。

对图 12-1 中各个组件进行一一介绍。
- Middleware：消息中间件，本例使用 RabbitMQ。
- Binder：消息中间件适配器，将中间件和 Stream 应用粘合起来，不同的中间件对应不同的 Binder。
- Channel：通道，应用通过一个明确的 Binder 与外界（中间件）通信。
- Application Core：Stream 自己实现的消息机制封装，包括分区、分组、发布订阅的语义，与具体中间件无关，这会让开发人员很容易地以相同的代码使用不同类型的中间件。

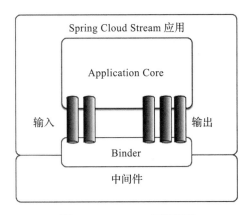

图 12-1　Stream 应用模型

12.2　示例

接下来的示例将以 RabbitMQ 为例。RabbitMQ 启动并指定容器端口 5672 映射到宿主机的 5672。

```
docker run -d -p5672:5672 rabbitmq
```

新建两个 Maven 项目，分别作用于消息的生产／发送和消费／接受，在 pom.xml 文件中新增如下片段，包括 Spring Cloud Stream 的核心依赖和针对 RabbitMQ 的实现依赖。

```
<dependencies>
    <dependency>
        <groupId>org.springframework.cloud</groupId>
        <artifactId>spring-cloud-starter-stream-rabbit</artifactId>
    </dependency>
    <dependency>
        <groupId>org.springframework.cloud</groupId>
        <artifactId>spring-cloud-stream</artifactId>
    </dependency>
</dependencies>
```

接下来在消息消费处理方 src/main/recource/Application.yml 配置文件中增加如下配置:

```yaml
spring:
  cloud:
    stream:
      bindings:
        input:             #channelName
          destination: mytopic #destination，可以认为是发布—订阅模型里面的topic
          binder: rabbit1
      binders:
        rabbit1:
          type: rabbit
          environment:
            spring:
              rabbitmq:
                host: localhost    #RabbitMQ服务器地址
                port: 5672
server:
  port: 8081
```

新建 StreamReceiverApplication.java 作为消息接收处理类:

```java
@SpringBootApplication
@EnableBinding(Sink.class)
@ComponentScan
public class StreamReceiverApplication {
    public static void main(String[] args) {
        org.springframework.context.ConfigurableApplicationContext context =
            SpringApplication.run(StreamReceiverApplication.class, args);
    }
    @StreamListener(Sink.INPUT)
    public void reader(String msg){
        System.out.println(msg);
    }
}
```

在消息生产方的配置文件中配置如下内容:

```yaml
spring:
  cloud:
    stream:
      bindings:
        input:
          destination: mytopic
          binder: rabbit1
      binders:
        rabbit1:
          type: rabbit
          environment:
            spring:
              rabbitmq:
```

```yaml
                host: localhost
                port: 5672
server:
    port: 8081
```

新建消息生产方类：

```java
@SpringBootApplication
@EnableBinding(Source.class)
@RestController
public class StreamSenderApplication {
    @Autowired
    StreamSenderApplication streamSenderApplication;
    @RequestMapping("/send")
    public String send(String msg){
        streamSenderApplication.sendMessage(msg);
        return "OK";
    }
    public static void main(String[] args) {
        SpringApplication.run(StreamSenderApplication.class, args);

    }
    @Autowired
    private Source source;
    //发送消息
    public String sendMessage(String msg) {
        try {
            source.output().send(MessageBuilder.withPayload(msg).build());
        } catch (Exception e) {
            e.printStackTrace();
        }
        return null;
    }
}
```

依次启动消费方应用、生产方应用。浏览器访问生产方提供的 HTTP 发送地址：

http://localhost:61221/send?msg=test

我们将会看到消费方的控制台输出出了我们从 URL 上发出的 test 的消息。

12.3 代码解析

首先看 StreamReceiverApplication 消息接收者的代码。类上标注了 EnableBinding 注解并传入了 Sink 类。@EnableBinding(Sink.class) 表示告知应用将此 Spring Boot 程序增加 Stream 通道监听功能，监听 org.springframework.cloud.stream.messaging.Sink 中名为 input 的输入管道。

```java
public interface Sink {
    String INPUT = "input";
```

```
    @Input(Sink.INPUT)
    SubscribableChannel input();
}
```

此外，Spring Cloud 还提供了输出管道 org.springframework.cloud.stream.messaging.Source、数据处理管道 org.springframework.cloud.stream.messaging.Processor，代码如下：

```
public interface Source {
    String OUTPUT = "output";
    @Output(Source.OUTPUT)
    MessageChannel output();
}
public interface Processor extends Source, Sink {
}
```

1. @Input、@Output

@Input、@Output 注解的入参为绑定的 BindingName，也可以根据需求自定义管道。

```
public interface Barista {
    @Input
    SubscribableChannel orders();
    @Output
    MessageChannel hotDrinks();
    @Output
    MessageChannel coldDrinks();
}
```

然后将其添加到 Spring Boot 程序主入口上的 @EnableBinding(value={ Barista.class，XXXX.class} 中，即可使 Spring 对其进行扫描监听。

Spring Framework 4 中包含了 spring-messaging 模块，它是从 Spring 集成的项目（如 Message/MessageChannel/MessageHandler 等）中提取出来的，这些被集成的项目是作为消息基础应用服务的。这个模块也包含了方法映射消息的注解的集合。

上述代码中 MessageChannel、SubscribableChannel 等类则是该模块中对消息进行处理抽象的类。

spring-messaging 提供了以下抽象对象，参见表 12-1。

表 12-1　spring-messaging 中的抽象对象

对　　象	说　　明
Message	一个带有头、载荷的消息
MessageHandler	处理消息的逻辑单元
MessageChannel	在发送者 / 接收者之间传输消息的信道的抽象，通道总是单向的
SubscribableChannel	继承 MessageChannel，用于传输消息到所有订阅者
ExecutorSubscribableChannel	继承 SubscribableChannel，使用异步线程池传输消息

2. 配置文件格式

我们可以定义多个 binding，分别为 binding 绑定相同或者不同的 Binder。

配置文件的格式为如下（其中含有 <> 的代码为自行定义）：

```
spring:
    cloud:
        stream:
            bindings:
                <channel-name1>:
                    destination: <topic-name> #可以认为是发布—订阅模型里面的topic
                    binder: <binder-name1>
                <channel-name2>:
                    destination: <topic-name> #可以认为是发布—订阅模型里面的topic
                    binder: <binder-name2>
            binders:
                <binder-name1>:
                    type: rabbit    #中间件类型 rabbit/redis/kafka……
                    environment:
                    #根据Binder类型所需要的配置项而有所不同
                        spring:
                            rabbitmq:
                                host: localhost    #RabbitMQ服务器地址
                                port: 5672
```

3. 对象注入

在 Spring Cloud Stream 的实例开发过程中，一般不需要接触处理过程中涉及的对象，只需要关注消息处理的业务代码模块即可，但是我们同样可以通过注入的方式获取到这些类。

比如，可以自行注入 Spring 创建的 Source 实例，然后通过 source.output().send (MessageBuilder.withPayload(name).build()) 来发送消息。

4. 聚合

Spring Cloud Stream 可以支持多种应用的聚合，可以实现多种应用输入/输出通道直接连接，而无须额外代价。其中支持如下聚合。

- source：带有名为 output 的单一输出通道的应用。典型情况下，该应用带有包含一个 org.springframework.cloud.stream.messaging.Source 的绑定。
- sink：带有名为 input 的单一输入通道的应用。典型情况下，该应用带有一个 org.springframework.cloud.stream.messaging.Sink 的绑定。
- processor：带有名为 input 的单一输入通道和带有名为 output 的单一输出通道的应用。典型情况下，该应用带有 org.springframework.cloud.stream.messaging.Processor. 的绑定。

可以通过创建一系列相互连接的应用将它们聚合到一起，其中，序列中一个元素的输出通道与下一个元素的输入通道连接在一起。序列可以由一个 Source 或者一个 Processor 开始，可以包含任意数目的 Processor，并由 Processor 或者 Sink 结束。

取决于开始和结束元素的特性，序列可以有一个或者多个可绑定的通道。

- 如果序列由 Source 开始、Sink 结束，应用之间直接通信并且不会绑定通道；
- 如果序列由 Processor 开始，它的输入通道会变成聚合的 input 通道并进行相应的绑定；
- 如果序列由 Processor 结束，它的输出通道会变成聚合的 output 通道并进行相应的绑定。

下面是使用 AggregateApplicationBuilder 功能类来实现聚合的示例，其为包含 Source、Processor 和 Sink 的工程。

```
@SpringBootApplication
@EnableBinding(Sink.class)
public class SinkApplication {
    private static Logger logger = LoggerFactory.getLogger(SinkModuleDefinition.class);
    @ServiceActivator(inputChannel=Sink.INPUT)
    public void loggerSink(Object payload) {
        logger.info("Received: " + payload);
    }
}
@SpringBootApplication
@EnableBinding(Processor.class)
public class ProcessorApplication {
    @Transformer
    public String loggerSink(String payload) {
        return payload.toUpperCase();
    }
}
@SpringBootApplication
@EnableBinding(Source.class)
public class SourceApplication {
    @Bean
    @InboundChannelAdapter(value = Source.OUTPUT)
    public String timerMessageSource() {
        return new SimpleDateFormat().format(new Date());
    }
}
```

每一个配置可用于运行一个独立的组件，在这个例子中，它们可以这样实现聚合：

```
@SpringBootApplication
public class SampleAggregateApplication {
    public static void main(String[] args) {
        new AggregateApplicationBuilder()
            .from(SourceApplication.class).args("--fixedDelay=5000")
            .via(ProcessorApplication.class)
            .to(SinkApplication.class).args("--debug=true").run(args);
    }
}
```

序列的开始组件被提供作为 from() 方法的参数，序列的结束组件被提供作为 to() 方法的参数，中间处理器组件则作为 via() 方法的参数。同一类型的多个 Processor 可以链接在一起（例如，可以使用不同配置的管道传输方式）。对于每一个组件，编译器可以为 Spring Boot 提供运行时参数。

12.4　Spring Integration 支持

1. 注解支持

因为 Spring Cloud Stream 本身就是基于 Spring Integration 的，所以在 Spring Cloud Stream 中同样可以支持并使用 Spring Integration 中的注解。

例如，可以使用 Spring Integration 中的 InboundChannelAdapter 来适配 Spring Cloud Stream 中的消息：

```
@EnableBinding(Source.class)
public class TimerSource {
    @Value("${format}")
    private String format;
    @Bean
    @InboundChannelAdapter(value = Source.OUTPUT, poller = @Poller(fixedDelay =
        "${fixedDelay}", maxMessagesPerPoll = "1"))
    public MessageSource<String> timerMessageSource() {
        return () -> new GenericMessage<>(now SimpleDateFormat(format).format(new
            Date()));
    }
}
```

又如，可以直接使用 @Transformer 来做消息的转换传递：

```
@EnableBinding(Processor.class)
public class TransformProcessor {
    @Transformer(inputChannel = Processor.INPUT, outputChannel = Processor.OUTPUT)
    public Object transform(String message) {
        return message.toUpperCase();
    }
}
```

Spring Cloud Stream 提供了 @StreamListener 注解，来覆盖 Spring Messaging 中原有的诸如 @MessageMapping、@JmsListener、@RabbitListener 之类的注解。@StreamListener 注解支持对接收的消息进行处理，特别是在处理特殊类型的结构化消息时更为方便。

2. 多函数消息处理

Spring Cloud Stream 支持将消息根据条件分发到多个 @StreamListener 的方法上。

这些方法必须满足如下条件：

- 无返回值；
- 单独只能处理一个消息。

下面是一个使用 @StreamListener 包含过滤条件的例子。在这个例子中，所有 Header 信息中 Type 带有值为 foo 的消息都将被调度到 receiveFoo 方法。

```
@EnableBinding(Sink.class)
@EnableAutoConfiguration
public static class TestPojoWithAnnotatedArguments {
    @StreamListener(target = Sink.INPUT, condition = "headers['type']=='foo'")
    public void receiveFoo(@Payload FooPojo fooPojo) {
    }
    @StreamListener(target = Sink.INPUT, condition = "headers['type']=='bar'")
    public void receiveBar(@Payload BarPojo barPojo) {
    }
}
```

12.5 Binder 解析

就像在 StreamSenderApplication 中，我们注入了 Source，并使用 source.output().send(MessageBuilder.withPayload(msg).build()) 来发送消息。Spring 会为每一个标注了 @Output @Input 的管道接口生成实现类。当 Spring 创建每个管道实现类的时候需要先创建对应的 Binder。

Spring Cloud Stream 的 Binder 接口定义了绑定消费方和绑定生产方的两个方法，并提供了默认的抽象实现 org.springframework.cloud.stream.binder.AbstractBinder。由各个消息中间件的实现模块来继承并实现 AbstractBinder，示例代码如下：

```
public interface Binder<T, C extends ConsumerProperties, P extends Producer Properties> {
    Binding<T> bindConsumer(String name, String group, T inboundBindTarget,
        C consumerProperties);
    Binding<T> bindProducer(String name, T outboundBindTarget, P producer Properties);
}
```

比如 RabbitMQ 对应 spring-cloud-starter-stream-rabbit 的实现类 org.springframework.cloud.stream.binder.rabbit RabbitMessageChannelBinder。

在 Binder 实现的依赖模块中，Spring Cloud Stream 约定在依赖 Jar 包中的 /META-INF/spring.binders 文件需要表明该 Binder 的 Type 以及配置相关的实现类。

spring-cloud-starter-stream-rabbit 中的 spring.binders 内容如下：

```
rabbit:\
org.springframework.cloud.stream.binder.rabbit.config.RabbitServiceAutoConfiguration
```

1. 动态绑定生产者

然而在实际的开发场景中，我们有时会需要动态地读取或发送至不同的 Topic，并且在运行时才能知道 Topic。针对这种场景，Spring Cloud Stream 也提供了支持。可以通过如下代码实现：

```
@EnableBinding
@Controller
@SpringBootApplication
public class SourceWithDynamicDestination {
    @Autowired
    private BinderAwareChannelResolver resolver;
    @RequestMapping("/bind")
    @ResponseBody
    public String handleRequest(@RequestParam String msg, @RequestParam String
        channelName, @RequestParam String contentType) {
        sendMessage(msg, channelName, contentType);
        return "OK";
    }
    private void sendMessage(String msg, String channelName, Object contentType) {
        MessageHeaders messageHeaders = new MessageHeaders(Collections.
            singletonMap(MessageHeaders.CONTENT_TYPE, contentType));
        Message<String> message = MessageBuilder.createMessage(msg,
            messageHeaders);
        resolver.resolveDestination(channelName).send(message);
    }
    public static void main(String[] args) {
    SpringApplication.run(SourceWithDynamicDestination.class, args);
    }
}
```

在上述代码中我们定义了一个 Controller 用于接收 Web 请求，传入三个参数。

- msg：消息内容；
- channelName：中间件 Topic；
- contentType：消息类型（如 application/json）。

我们定义了 sendMessage 方法，使用注入的 BinderAwareChannelResolver 来针对动态的 Topic 发送消息。

我们尝试发起 HTTP 请求：

```
http://localhost:8081/bind?msg= mytestmsg&channelName=mytopic&contentType=
    application/json
```

大家可以从各自的 StreamReceiverApplication 控制台看到我们向 mytopic 发送的 mytest。

2. 消费者分组

在一些场景下，某个消息只能被一个消费者所消费，但是在实际的部署中，可能这个消费者部署了多台实例想共享消费这个消息。Spring Cloud Stream 利用消费者组定义这种行为（这种分组类似于 Kafka 的消费组），每个消费者通过 spring.cloud.stream.bindings.input.group 指定一个组名称，以图 12-2 所示的消费者为例，应分别设置 spring.cloud.stream.bindings.input.group=hdfsWrite 和 spring.cloud.stream.bindings.input.group=average。

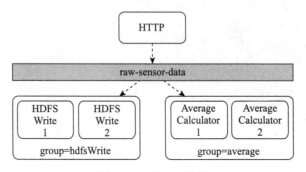

图 12-2　消费者示例

所有订阅指定 Topic 的组都会收到发布数据的一份副本，但是每一个组内只有一个成员会收到该消息。默认情况下，当一个组没有指定时，Spring Cloud Stream 将分配给一个匿名的、独立的只有一个成员的消费组，该组与所有其他组都处于一个发布－订阅关系中。

3. 持久性

Spring Cloud Stream 一致性模型中，消费者组订阅是持久的，一旦组中至少有一个成员创建了订阅，这个组就会收到消息，即使组中所有的应用都被停止了，组仍然会收到消息。

然而匿名订阅是非持久的，一些 Binder 的实现（如 RabbitMQ）可以创建非持久化组订阅。

在一般情况下，将应用绑定到指定 Topic 时，最好指定一个消费者组，当扩展一个 Spring Cloud Stream 应用时，必须为每个输入 binding 指定一个消费组，这可以防止应用的实例接收重复的消息。

4. 分区支持

Spring Cloud Stream 支持在一个应用的多个实例之间的数据分区，在分区的情况下，物理通信介质（如 Topic 代理）被视为多分区结构。一个或多个生产者应用实例将数据发送给多个消费者应用实例，并保证有共同特性的数据由相同的消费者实例处理。

Spring Cloud Stream 提供了一个通用的抽象，用于以统一方式进行分区处理，因此分区可以用于自带分区的代理（如 Kafka），或者不带分区的代理（如 RabbitMQ），如图 12-3 所示。

分区在有状态处理中是一个很重要的概

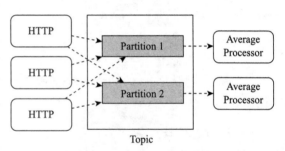

图 12-3　分片支持示意图

念，其重要性体现在性能与一致与上，要确保所有相关数据被一并处理。

12.6　常用配置

1. Spring Cloud Stream 配置项

以 spring.cloud.stream 为前缀的配置，有如下可选项。

- instanceCount：集群模式中实例数量，如果使用 Kafka 则必须配置该项。默认值为 1。
- instanceIndex：集群模式中该实例在集群中的编号，取值范围为 0~instanceCount-1。
- dynamicDestinations：Topic 列表，如果设置了该选项，那么只有出现在该列表中的 Topic 才能被关联。默认值为空（允许关联任何 Topic）。
- defaultBinder：默认使用哪个 Binder。默认值为空。

2. Binder 配置项

以 spring.cloud.stream.binders.<binder-name> 为前缀的配置，有如下可选项。

- type：Binder 类型，对应 /META-INF/spring.binders 中的 Type。
- inheritEnvironment：是否允许从程序环境变量中继承配置项，默认值为 True。
- environment：该类型的 Binder 所需要的特定配置项。
- defaultCandidate：是否可以被当作默认 Binder，默认值为 true。

3. Binding 配置项

以 spring.cloud.stream.bindings.<channelName> 为前缀的配置项，有如下可选项。

- Destination：消息中间件的 Topic。
- Group：消费者分组，下文会介绍关于消费者分组支持的内容，默认值为空。
- contentType：消息类型，用于 Spring Cloud Stream 对消息内容的自动转换，下文会详细介绍。
- Binder：该 binding 绑定的对应 Binder，默认值为空。在为空的情况下，如果有默认 Binder 则选取默认 Binder。

4. Binding 中消费者配置项

以 spring.cloud.stream.bindings.<channelName>.consumer 为前缀的配置项，有如下可选项。

- Concurrency：支持的消费者并发数量，默认值 1。
- Partitioned：消费者是否接收从分片的生产者的数据，默认值 false。
- maxAttempts：最大失败重试次数，1 表示不重试，默认值为 3。
- backOffInitialInterval：重试间隔时间（ms），默认值 1000。

- backOffMaxInterval：最大重试间隔时间（ms），默认值 10000。
- instanceIndex：当前实例在集群中的序号，默认值 −1。
- instanceCount：当前实例集群总数，默认值 −1。

12.7 小结

当我们有了 Spring Cloud Stream，开始以消息驱动的业务场景实现时，可以不用太关注具体的消息中间件具体实现。我们直接面向 Spring Cloud Stream 开发时屏蔽了中间件的实现对具体开发产生的差异性影响，做到了业务开发与工具实现的彻底解耦，好处不言而喻。

下一章将学习使用 Spring Cloud Bus 如何进行事件广播。

第 13 章　Chapter 13

消息总线：Spring Cloud Bus

Spring Cloud Bus 通过轻量级消息代理连接各个分布的节点，以广播状态的变化（如配置变化）或者其他的消息指令。Spring Cloud Bus 的一个核心思想是通过分布式的启动器对 Spring Boot 应用进行扩展，也可以用来建立多个应用之间的通信频道。目前唯一实现的方式是用 AMQP 消息代理作为通道队列，目前常用的有 Kafka 和 RabbitMQ。利用 Bus 机制可以做很多的事情，其中配置中心客户端刷新就是典型的应用场景之一。当 Git Repository 改变时，通过 POST 请求 Config Server 的 /bus/refresh，Config Server 会从 Repository 获取最新的信息并通过 AMQP 或者 Kafka 等传递给 Client，如图 13-1 所示。

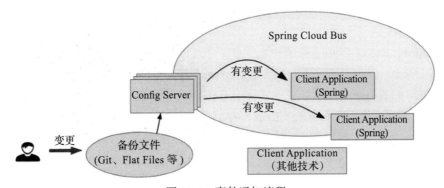

图 13-1　事件通知流程

Spring Cloud Bus 的更新只对 3 种情况有效：
- @ConfigurationProperties；
- @RefreshScope；
- 日志级别。

13.1 使用 Spring Cloud Bus

接下来以 RabbitMQ 为例，我们在 3.3 节的项目中的 pom 文件中添加对 Spring Cloud Bus 和 Actuator 的依赖：

```
<dependency>
    <groupId>org.springframework.cloud</groupId>
    <artifactId>spring-cloud-starter-bus-amqp</artifactId>
<dependency>
    <groupId>org.springframework.cloud</groupId>
    <artifactId>spring-cloud-starter-bus </artifactId>
</dependency>
<dependency>
    <groupId>org.springframework.boot</groupId>
    <artifactId>spring-boot-starter-actuator</artifactId>
</dependency>
```

在配置文件中配置 RabbitMQ 的连接信息，如果使用 Kafka，则添加 spring-cloud-starter-bus-Kafka 的依赖即可，在 application.yml 进行如下配置：

```
spring:
    rabbitmq:
        host: mybroker.com
        port: 5672
        username: user
        password: secret
```

需要启动 2 个或者 2 个以上的 ConfigClient1 和 ConfigClient2。读者可以自行修改端口启动多个 Config Client。

然后我们尝试改动 Git 仓库中 Master 分支下的 configServerDemo-test.properties 的 key1 的 Value 值。

根据 Spring Cloud Config 章节中的测试，我们得知需要请求 ConfigClient1 的 /refresh 端点，告知 ConfigClient1 服务器去 Config Server 重新获取节点。

然而在集成 Spring Cloud Bus 后，我们只需要通知连接在同一个 RabbitMQ 下的任意 Config Client，即可使所有 Config Client 都接到通知并去 Config Server 中获取最新的配置。

13.2 进阶场景

1. Git 的 WebHook 通知

在实际的使用过程中，我们不可能每次修改完配置之后，手动请求 ClientServer 的 /refresh 端点。此时，我们可以利用 Git 提供的 WebHook 功能。

WebHook 允许第三方应用监听 GitHub 上的特定事件，在这些事件发生时通过 HTTP POST 方式通知（超时 5s）到第三方应用指定的 Web URL。例如项目有新的内容 Push，或

Merge Request 有更新等。WebHook 可方便用户实现自动部署、自动测试、自动打包、监控项目变化等。

比如在图 13-2 实例中，我们可以将配置仓库的 WebHook 配置为内容发生改变时，请求 /refresh 端点。那么这样是否可以达到任何配置文件的改动立马通过 Bus 通知到所有 Client 重新更新配置的效果呢？

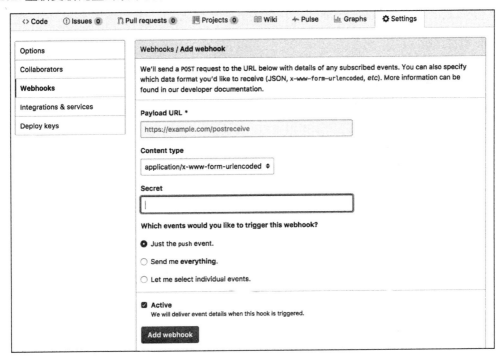

图 13-2　WebHook 配置界面

2. 局部刷新

通过 /bus/refresh 的 destination 参数我们可以指定刷新某一台 Config Client 实例，例如 /bus/refresh?destination=cloudConfigDemo:7004。其中 destination 为 Spring 的 ApplicationContext 的 ID。可以通过注入 ApplicationContext，调用其 getId() 方法获取。默认 ID 是由 spring.application.name:server.port 拼接而成。/bus/refresh?destination= cloudConfigDemo:** 则将会让所有 cloudConfigDemo 服务的实例进行刷新。如果 Config Client 部署在不同的物理服务器上，可能端口也是一样的，那么默认生成的 ApplicationContextID 也是一样的。这种情况下我们可以通过 spring.application.index 指定实例的序号来避免 ApplicationContextID 的情况。

```
spring.application.index=${INSTANCE_INDEX}。
```

3. 消息中间件配置

Spring Cloud Bus 使用 Spring Cloud Stream 来进行消息广播。Bus 提供了方便的 starter 来适配 AMQP（RabbitMQ）和 Kafka（spring-cloud-starter-bus-[amqp,kafka]）。

例如：对于 AMQP 代理的地址可以通过 spring.rabbitmq.* 等配置项进行设置。Spring Cloud Bus 有一些自己的配置项在 spring.cloud.bus.* 中（如 spring.cloud.bus.destination 可以指定 topic 的名字），一般用默认的即可。更多消息代理的设置，可以参考第 10 章。

4. 事件追踪

在 Spring Cloud Bus 内部，配置刷新通知等功能都是基于 Spring 的事件机制来实现的。这些事件（都继承于 RemoteApplicationEvent）都是可被追踪的，只需要设置 spring.cloud.bus.trace.enabled=true。配置后，Spring 将事件信息都以 List 的形式存储在 TraceRepository 中，我们可以看到每一个事件的发送以及每一个服务实例的确认信息。例如，通过 HTTP/trace 端点：

```
{
    "timestamp": 1494166629048,
    "info": {
        "signal": "spring.cloud.bus.ack",
        "event": "RefreshRemoteApplicationEvent",
        "id": "c96816f7-e909-4003-9887-65b2a2fc1588",
        "origin": "cloudConfigDemo7004:dev:7004",
        "destination": "cloudConfigDemo7004:dev:7004"
    }
},
{
    "timestamp": 1494166628240,
    "info": {
        "method": "GET",
        "path": "/configServerDemo/dev/master",
        "headers": {
            "request": {
                "accept": "application/json, application/*+json",
                "user-agent": "Java/1.8.0_131",
                "host": "192.168.1.5:8888",
                "connection": "keep-alive"
            },
            "response": {
                "X-Application-Context": "myConfigServer:8888",
                "Content-Type": "application/json;charset=UTF-8",
                "Transfer-Encoding": "chunked",
                "Date": "Sun, 07 May 2017 14:17:08 GMT",
                "status": "200"
            }
        },
        "timeTaken": "997"
    }
},
```

用户也可以自己添加 @EventListener 来处理 Ack 信号，对于应用来说，有两种事件类型：AckRemoteApplicationEvent 和 SentApplicationEvent。还可以直接获得 TraceRepository 并从中得到数据。

5. 广播自定义事件

Spring Cloud Bus 可以携带传输任何类型的 RemoteApplicationEvent 事件，默认传输是通过 JSON 方式，如果需要自定义注册一个新类型事件，则默认将其放在 org.springframework.cloud.bus.event 包中。

你可以在自定义类上使用 @JsonTypeName 自定义事件名称。注意，生产者和消费者都需要访问类定义。

如果不想将自定义事件类放在 org.springframework.cloud.bus.event 包下，则必须使用 @RemoteApplicationEventScan 指定要扫描 RemoteApplicationEvent 类型事件包。使用 @RemoteApplicationEventScan 指定包含子包。

例如，用户有一个名为 FooEvent 的自定义事件：

```
package com.acme;
public class FooEvent extends RemoteApplicationEvent {
    ...
}
```

可以通过以下方式向解析器注册此事件：

```
package com.acme;
@Configuration
@RemoteApplicationEventScan
public class BusConfiguration {
    ...
}
```

不指定值，将注册使用 @RemoteApplicationEventScan 注解的类包。在本例中，com.acme 将使用 BusConfiguration 的包进行注册。

你还可以使用 @RemoteApplicationEventScan 上的值，basePackages 或 basePackageClasses 属性显式指定要扫描的包。例如：

```
package com.acme;
@Configuration
//@RemoteApplicationEventScan({"com.acme", "foo.bar"})
//@RemoteApplicationEventScan(basePackages = {"com.acme", "foo.bar", "fizz.buzz"})
@RemoteApplicationEventScan(basePackageClasses = BusConfiguration.class)
public class BusConfiguration {
    ...
}
```

上面的 @RemoteApplicationEventScan 的所有示例都是等效的，因为 com.acme 包将通过在 @RemoteApplicationEventScan 上显式指定包来注册。

> **注意** 你可以指定要扫描的多个包。

13.3 小结

由于消息总线在微服务架构系统的广泛使用，所以它同配置中心一样，几乎是微服务架构中的必备组件。Spring Cloud Bus 作为 Spring Cloud 中控制消息广播的组件，通过它可以非常容易地搭建起消息总线，同时实现一些消息总线中的常用功能。

第 14 章 批处理：Spring Cloud Task

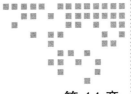

Spring Cloud Task 是支持短生命周期的微服务 Spring Cloud 子项目，作用于定时任务、批处理一类的需求场景。Spring Cloud Task 提供如下几点功能：

- 任务执行日志记录（数据库）；
- 任务的切面配置；
- Spring Batch 的集成；
- 远程 Task 获取执行（如从 Nexus 私服获取）；
- 资源编排平台的支持（YARN、Kubernetes 等）；
- Spring Cloud Stream 的集成。

14.1 使用 Spring Cloud Task

我们先在 IDE 中创建以 Maven 作为依赖管理的项目，并在 pom.xml 中添加如下依赖：

```
<dependency>
    <groupId>org.springframework.boot</groupId>
    <artifactId>spring-boot-starter</artifactId>
</dependency>
<dependency>
    <groupId>org.springframework.cloud</groupId>
    <artifactId>spring-cloud-task-core</artifactId>
    <version>1.0.0.RELEASE</version>
</dependency>
</dependencies><build><plugins><plugin>
    <groupId>org.springframework.boot</groupId>
    <artifactId>spring-boot-maven-plugin</artifactId>
</plugin></plugins></build></project>
```

其中 spring-cloud-task-core 的依赖为 Spring Cloud Task 的核心依赖包。后续我们还会加入一些可选的依赖包，如 spring-boot-starter-jdbc。

接下来在项目的 src/main/java 中创建第一个类：

```
@SpringBootApplication
@EnableTask
public class SampleTask {
    @Bean
    public CommandLineRunner commandLineRunner() {
        return new HelloWorldCommandLineRunner();
    }
    public stati cvoid main(String[] args) {
        SpringApplication.run(SampleTask.class, args);
    }
    public static class HelloWorldCommandLineRunner implements CommandLineRunner {
        @Override
        publi cvoid run(String... strings) throws Exception {
            System.out.println("Hello World!");
        }
    }
}
```

在上述 SampleTask 类中，@EnableTask 表示通知 Spring 开启任务扫描功能，并定义了一个 CommandLineRunner 接口的实现类 HelloWorldCommandLineRunner 作为被 Spring 管理的 Bean。在 HelloWorldCommandLineRunner 的 Run 方法中打印输出"Hello World!"。

在 src/main/resources 中新建 application.properties 并添加如下内容：

```
logging.level.org.springframework.cloud.task=DEBUG
spring.application.name=helloWorld
```

运行之后控制台打印出的结果包含如下几行日志：

```
Creating: TaskExecution{executionId=0, exitCode=null, taskName='helloWorld',
    startTime=Tue Mar 07 10:58:02 CST 2017, endTime=null, exitMessage='null',
    arguments=[]}
Hello World!
Updating: TaskExecution with executionId=0 with the following {exitCode=0,
    endTime=Tue Mar 07 10:58:02 CST 2017, exitMessage='null'}
```

其中第一行表示创建了该任务日志，第二行是运行 Task 内容打印，第三行更新任务。

14.2 进阶场景

14.2.1 数据库集成

Spring Cloud Task 默认将 Task 生命周期记录在内存中的 Map 中，在生产环境使用的时候，我们可能需要将其存储到数据库中。Spring Cloud Task 也提供了极其简便快捷的方法，给予支持。

目前支持的数据库如下：
- DB2
- H2
- HSQLDB
- MySQL
- Oracle
- Postgres
- SQLServer

接下来将以 MySQL 为例。

在项目的 Maven pom.xml 中添加对数据库的依赖：

```
<dependency>
    <groupId>org.springframework.boot</groupId>
    <artifactId>spring-boot-starter-jdbc</artifactId>
</dependency>
<dependency>
    <groupId>mysql</groupId>
    <artifactId>mysql-connector-java</artifactId>
    <version>5.1.39</version>
</dependency>
```

在 Application.properties 中添加数据库的连接信息配置：

```
spring.datasource.url=jdbc:mysql://localhost:3306/test
spring.datasource.username=root
spring.datasource.password=root
spring.datasource.driver-class-name=com.mysql.jdbc.Driver
```

接下来启动 Application，将看到日志信息中输出了带有如下内容的日志：

```
Initializing task schema for mysql database
Executing SQL script from class path resource [org/springframework/cloud/task/
    schema-mysql.sql]
Executed SQL script from class path resource [org/springframework/cloud/task/
    schema-mysql.sql] in 704 ms.
```

Spring Cloud Task 为我们自动创建了数据库表，并将本次执行的日志存储在内了。

图 14-1 所示为 Task 的表结构的 Schema。

其中每个表的作用如下。
- TASK_EXECUTION：任务执行记录。
- TASK_EXECUTION_PARAMS：任务执行参数。
- TASK_SEQ：任务 ID 生成器。
- TASK_TASK_BATCH：SpringBatch 步骤记录。

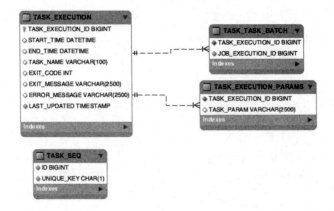

图 14-1　Task 的表结构

14.2.2　任务事件监听

Spring Cloud Task 提供了两种方式处理任务的启动、结束、异常这三种事件。

1. 接口式

我们可以实现 TaskExecutionListener 接口，来实现对任务的事件监听。

```
TaskExecutionListener
public interface TaskExecutionListener {
    public void onTaskStartup(TaskExecution taskExecution);
    public void onTaskEnd(TaskExecution taskExecution);
    public void onTaskFailed(TaskExecution taskExecution, Throwable throwable);
}
```

其中每个方法的作用如下。

- onTaskStartup：任务的启动事件监听。
- onTaskEnd：任务的结束事件监听。
- onTaskFailed：任务的异常事件监听。

新建一个 TaskExecutionListener 的实现类 MyTaskExecutionListener：

```
public class MyTaskExecutionListener implements TaskExecutionListener {
    @Override
    public void onTaskStartup(TaskExecution taskExecution) {
        System.out.println("Before task:"+taskExecution.getTaskName());
    }
    @Override
    public void onTaskEnd(TaskExecution taskExecution) {
        System.out.println("After task:"+taskExecution.getTaskName());
    }
    @Override
    public void onTaskFailed(TaskExecution taskExecution, Throwable throwable) {
```

```
            System.out.println("Failed task:"+taskExecution.getTaskName()+","+throwable.
                getMessage());
    }
}
```

在之前的 SampleTask 类中新增如下代码，定义 MyTaskExecutionListener 为 SpringBean：

```
@Bean
public MyTaskExecutionListener myBean() {
    return new MyTaskExecutionListener();
}
```

尝试运行 SampleTask 类，将会看到之前打印 HelloWorld 的前后多了事件监听的输出：

```
Before task:helloWorld
Hello World!
After task:helloWorld
```

2. 注解式

Spring Cloud Task 同样提供了相比 TaskExecutionListener 更为方便的注解式来实现事件的任务监听，如下罗列了提供的注解列表和对应的作用。

- @BeforeTask：对应 TaskExecutionListener.onTaskStartup。
- @AfterTask：对应 TaskExecutionListener.onTaskEnd。
- @FailedTask：对应 TaskExecutionListener.onTaskFailed。

将 MyTaskExecutionListener 去掉，将 TaskExecutionListener 的接口实现修改为如下代码，效果与实现 TaskExecutionListener 接口等同。

```
public class MyTaskExecutionListener  {
    @BeforeTask
    public void onTaskStartup(TaskExecution taskExecution) {
        System.out.println("Before task:"+taskExecution.getTaskName());
    }
    @AfterTask
    public void onTaskEnd(TaskExecution taskExecution) {
        System.out.println("After task:"+taskExecution.getTaskName());
    }
    @FailedTask
    public void onTaskFailed(TaskExecution taskExecution, Throwable throwable) {
        System.out.println("Failed task:"+taskExecution.getTaskName()+",
            "+throwable.getMessage());
    }
}
```

14.2.3 相关配置项

- spring.cloud.task.tablePrefix=<yourPrefix>：自动创建的表名默认是使用 TASK_（下划线）做前缀，我们可以通过该配置指定、表明前缀。不过需要注意的是，此配置对初始化创建表不生效。
- spring.cloud.task.initialize.enable=<true or false>：是否允许启动时自动初始化表结构。
- spring.cloud.task. executionid=< yourtaskId >：自定义 ID，存储在 TASK_EXECUTION 表的 EXECUTION_ID 字段。
- spring.cloud.task.name=<taskname>：默认情况下，SimpleTaskNameResolver 会按照优先级读取两个配置：spring.cloud.task.name 和 ApplicationContext.getId()。

12.4.4 整合 Spring Cloud Stream

Task 本身是可以单独执行的，但集成到一个更大的生态系统，可以使它用于更复杂的处理和协调。本部分介绍了 Spring Cloud Task 和 Spring Cloud Stream 的集成。

Spring 允许用户从 Stream 中启动 Task。这是通过创建一个 Sink 来监听包含 TaskLaunchRequest 的消息来完成的。TaskLaunchRequest 包含要执行的任务的 Maven 坐标。例如，可以创建一个 Stream，该 Stream 包含一个 processor，能够从 httpSource 接收数据并创建一个包含 TaskLaunchRequest 的 GenericMessage，然后 TaskSink 将接收到 GenericMessage 并启动 Task。

要创建 taskSink，用户只需要创建一个包含 EnableTaskLauncher 注解的 Spring Boot 应用程序。例如：

```
@SpringBootApplication
@EnableTaskLauncher
public class TaskSinkApplication {
    public static void main(String [] args){
        SpringApplication.run(TaskSinkApplication.class, args);
    }
}
```

14.3 源码解析

1. Runner 接口

Spring Cloud Task 提供了两种接口的 Task：CommandLineRunner 和 ApplicationRunner。当标注 @EnableTask 时，Spring Boot 将会执行实现这两个接口的 Bean。

```
public interface CommandLineRunner {
    void run(String... args) throws Exception;
}
public interface ApplicationRunner {
    void run(ApplicationArguments args) throws Exception;
}
```

ApplicationRunner 中 run 方法的参数为 ApplicationArguments，而 CommandLineRunner

接口中 run 方法的参数为 String 数组。当只需要见到简单的命令行传入参数时，可以使用 CommandLineRunner。而当需要复杂的传入参数形式和详细信息时，就需要使用 ApplicationRunner 了。

2. @EnableTask 注解

```
@Target(ElementType.TYPE)
@Retention(RetentionPolicy.RUNTIME)
@Documented
@Inherited
@Import(SimpleTaskConfiguration.class)
public @interface EnableTask {
}
```

@EnableTask 注解被标注时，程序将开启 Spring Cloud Task 的功能支持。通过 @EnableTask 的源码可以看到，在默认情况下会启用 SimpleTaskConfiguration 类作为任务配置类。其中定义了如下属性：

```
@Configuration
@EnableTransactionManagement
public class SimpleTaskConfiguration {
    protected static final Log logger = LogFactory.getLog(SimpleTaskConfiguration.class);
    @Autowired(required = false)
    private Collection<DataSource> dataSources;
    @Autowired
    private ConfigurableApplicationContext context;
    @Autowired(required = false)
    private ApplicationArguments applicationArguments;
    private boolean initialized = false;
    private TaskRepository taskRepository;
    private TaskLifecycleListener taskLifecycleListener;
    private TaskListenerExecutorFactoryBean taskListenerExecutorFactoryBean;
    private PlatformTransactionManager platformTransactionManager;
    private TaskExplorer taskExplorer;
}
```

3. skLifecycleListener

其中 TaskLifecycleListener 类为 Spring 事件监听实现类，Application 启动时触发 doTaskStart()：

```
private void doTaskStart() {
    if(!started) {
        List<String> args = new ArrayList<>(0);
        if(this.applicationArguments != null) {
            args = Arrays.asList(this.applicationArguments.getSourceArgs());
        }
        this.taskExecution = this.taskRepository.createTaskExecution(this.taskNameResolver.
            getTaskName(), new Date(), args);
```

```
        }
        else {
            logger.error("Multiple start events have been received.  The first
                one was " +
                    "recorded.");
        }
        taskExecution.setExitMessage(invokeOnTaskStartup(taskExecution).
            getExitMessage());
}
```

调用 TaskRepository 的 createTaskExecution 方法创建任务生命周期日志。

4. TaskRepository

TaskRepository 类为 Task 执行日志记录的操作类，默认实现为内存 ConcurrentMap 数据库记录。后续将讲解如何实现数据库记录的配置。createTaskExecution 返回 TaskExecution 类记录 Task 的详细信息。

TaskExecution 结构如下。

- Executionid：任务 ID。
- exitCode：返回编码，通过 SpringEvent 的终止事件得到，使用 ExitCodeExceptionMapper 的实现类映射得出。
- taskName：任务名称，通过 TaskNameResolver 获取，默认返回 spring.cloud.task.name 配置项的值，为空时返回 ID。
- startTime：任务开始事件，以 SmartLifecycle 的 start 方法启动事件为准。
- endTime：任务结束事件，通过监听 ApplicationReadyEvent 获得。
- exitMessage：任务结束的返回值，在 TaskExecutionListener 的 onTaskEnd 方法中填入。
- errorMessage：错误信息，如果任务发生异常，通过 ApplicationFailedEvent 事件监听获得。

14.4　小结

Spring Cloud Task 有助于更简单地创建短时运行的微服务。Spring Cloud 团队为在 JVM 上运行短时应用的需求提供基本的技术支持，由它开发的项目可以在本地云或者 Spring Data Flow 中启动。通过本章的学习，可以理解 Spring Cloud Task 的概况、用途和使用方法，以及 Spring Cloud Task 的设计理念。

第三篇 *Part 3*

微服务实战篇

前两篇我们已经把 Spring Cloud 生态中的每一个组件都进行了梳理。本篇中将针对之前学习的各个组件进行综合实战，通过案例让大家更深刻地理解每个组件。我们还会介绍微服务体系下的持续性构建交付的理论，阐述在软件研发生产线上的流程工具。

第 15 章

利用 Docker 进行编排与整合

基于前面讲述的内容，我们将 Spring Cloud 的各个组件整合成一个五脏俱全的小项目，通过 Docker Compose 进行编排、整合，达到实战练习的目的。

15.1 Docker 基础应用

Docker 使用 Google 公司推出的 Go 语言进行开发，基于 Linux 内核的 cgroup、namespace，以及 AUFS 类的 Union FS 等技术，对进程进行封装隔离，属于操作系统层面的虚拟化技术。由于隔离的进程独立于宿主和其他隔离的进程，因此也称其为容器。最初实现是基于 LXC，从 0.7 版本以后开始去除 LXC，转而使用自行开发的 libcontainer，从 1.11 版本开始，则进一步演进为使用 runC 和 containerd。

Docker 在容器的基础上，从文件系统、网络互联到进程隔离等进行了进一步的封装，极大地简化了容器的创建和维护，这使其比虚拟机技术更为轻便、快捷。

> **Docker 安装**：具体安装细节可以参照 https://docs.docker.com。Docker 原本是在 Linux 上开发的，Windows 下则采用 Docker Machine 的方式，建一个虚拟机，在虚拟机里面运行 Docker。本来 Mac 也是采用虚拟机的方式，后来有了 Docker for Mac，采用 HyperKit——一种更轻量级的虚拟化方式。当然，还可以用 docker machine 的方式运行。

15.1.1 Docker 基础

要弄清楚两个概念：容器与镜像。以装机为例，容器就是装完后的系统，镜像就是 ISO 安装盘，即系统镜像。

我们在用 Git 的时候，需要涉及 pull、commit、push 等操作。先拉取代码（pull），然后在一个基础的文本上不断地增加（commit），像积木一层层叠加，累积上去，最后再推送（push）上去。Docker 也是采用这种形式。

下面简单介绍一下 Docker 命令。

- docker version / docker info：查看基本信息，遇到使用问题或者 bug，可以到社区里报告。
- docker pull：拉取镜像。
- docker run：从镜像创建一个容器并运行指定的命令，常用参数为 -i、-d，建议用 -name 命名容器，命名后可以使用容器名访问它。
- docker exec -ti CONTAINER /bin/bash：连接到容器上运行 bash。
- docker logs CONTAINER：查看日志，如 run 命令后的运行结果，docker logs -f 用于查看实时的日志。
- docker kill：杀死 Docker 容器进程。可以使用 docker kill $(Docker ps -aq) 杀死所有的 Docker 进程，后者打印出所有容器的容器 ID（包括正在运行的和没有运行的）。
- docker rm CONTAINER：删除一个容器，记得要先停止正在运行的容器，再去删除它。
- docker exec -itbash -c 'cat > /path/to/container/file' < /path/to/host/file：容器外向容器内复制文件（也可以用挂载的形式）。
- docker commit -a "mike" -m " 镜像的一些改动 " CONTAINER：在容器内做了某种操作后，如增加了一个文件，可以用这个命令提交修改，重新打包为镜像。
- docker push：推送镜像。
- docker history IMAGES：查看镜像的修改历史。
- docker ps -a | grep "Exited" | awk '{print $1 }'| xargs docker rm：删除已停止退出的容器。
- docker rmi $(Docker images | awk '/^/ {print $3}')：删除 tag 为 NONE 的容器。

15.1.2　Dockerfile 基础

Dockerfile 是一个文本文件，包含一条条的指令（instruction），每一条指令构建一层，因此每一条指令的内容就是描述该层应当如何构建。由于构建 Docker 镜像时需要用到这个文件，所以先来简单学习一下 Dockerfile 的关键字和语法。

- FROM：指定所创建镜像的基础镜像。格式为 FROM <image> 或 FROM <image>:<tag>。第一条指令必须为 FROM 指令。
- MAINTAINER：作者签名，格式为 MAINTAINER <name>。
- ENV：配置环境变量。格式为 ENV <key><value>。指定一个环境变量，这个变量会被后续 RUN 指令使用，并在容器运行时保持。例如，ENV PATH /usr/local/

nginx/sbin:$PATH。
- COPY：将本地主机的 <scr> 路径下的内容复制到容器中的 <dest> 路径下，一般情况下推荐使用 COPY 而不是 ADD。
- EXPOSE：暴露容器的端口。格式为 EXPOSE <port> [<port>...]，告诉 Docker 服务端容器暴露的端口号，供外部调用方使用。在启动容器时需要通过 -P 参数标识 Docker 自动分配一个外部端口将传输数据转发到指定的端口。
- WORKDIR：后续命令的执行路径。格式为 WORKDIR /path/to/workdir。
- RUN：执行相应的命令，这是在容器构建这一步中执行的命令，一般用作安装软件，操作的结果将持久化保存在容器里面。格式为 RUN <command> 或 RUN ["executable", "param1", "param2"]。
- ENTRYPOINT：配置容器启动后执行的命令，并且不可被 Docker run 提供的参数覆盖。每个 Dockerfile 中只能有一个 ENTRYPOINT，当指定多个时，只有最后一个生效。

有两种格式：

ENTRYPOINT ["executable", "param1", "param2"]；

ENTRYPOINT command param1 param2（Shell 中执行）。

- VOLUME：挂载目录，格式为 VOLUME ["/data"]，创建一个可以从本地主机或其他容器挂载的挂载点，一般用来存放数据库和需要保持的数据等。
- ADD：相当于 COPY，但是比 COPY 功能更强大，格式为 ADD <src><dest>。该命令将复制指定的文件 <src> 到容器中的 <dest> 路径。其中 <src> 参数可以是 Dockerfile 所在目录的一个相对路径，也可以是一个 URL，还可以是一个 tar 文件，复制进容器会自动解压。
- USER：格式为 USER daemon。指定运行容器时的用户名或 UID，后续的 RUN 也会使用指定用户。当服务不需要管理员权限时，可以通过该命令指定运行用户，并且可以在之前创建所需要的用户，例如：RUN useradd -s /sbin/nologin -M www。
- CMD：指定启动容器时执行的命令，每个 Dockerfile 只能有一条 CMD 命令。如果指定了多条命令，只有最后一条会被执行。如果用户启动容器时指定了运行的命令，则会覆盖 CMD 指定的命令。

支持三种格式：

CMD ["executable","param1","param2"]：使用 exec 执行，推荐方式；

CMD command param1 param2：在 /bin/bash 中执行，提供给需要交互的应用；

CMD ["param1","param2"]：提供给 ENTRYPOINT 的默认参数。

- ONBUILD：在构建本镜像时不生效，在基于此镜像构建镜像时生效。格式为 ONBUILD [INSTRUCTION]，用于配置当所创建的镜像作为其他新创建镜像的基础镜像时所执行的操作指令。

ENTRYPOINT 和 CMD 的区别：ENTRYPOINT 指定了该镜像启动时的入口，CMD 则指定了容器启动时的命令，当两者共用时，完整的启动命令如 ENTRYPOINT + CMD。使用 ENTRYPOINT 的好处是，启动镜像就像启动了一个可执行程序，在 CMD 上仅需要指定参数；另外，在需要自定义 CMD 时不容易出错。

> 提示　每次执行 RUN 的时候都是在默认路径执行的，如果要到固定路径下执行命令，应在之前加 WORKDIR，或者使用 RUN(cd workpath && echo "mike")，这样把 cd 命令与相应的执行命令用括号括了起来。

一般情况下，Dockerfile 分为四部分：基础镜像信息、维护者信息、镜像操作指令和容器启动时执行指令。# 符号为 Dockerfile 中的注释。先看下面的小例子：

```
# This my first nginx Dockerfile
# Version 1.0
# Base images基础镜像
FROM centos
#MAINTAINER维护者信息
MAINTAINER hujinhan
#ENV设置环境变量
ENV PATH /usr/local/nginx/sbin:$PATH
#ADD 文件放在当前目录下，复制时会自动解压
ADD nginx-1.8.0.tar.gz /usr/local/
ADD epel-release-latest-7.noarch.rpm /usr/local/
#RUN执行以下命令
RUN rpm -ivh /usr/local/epel-release-latest-7.noarch.rpm
RUN yum install -y wget lftp gcc gcc-c++ make openssl-devel pcre-devel pcre && 
    yum clean all
RUN useradd -s /sbin/nologin -M www
#WORKDIR相当于cd
WORKDIR /usr/local/nginx-1.8.0
RUN ./configure --prefix=/usr/local/nginx --user=www --group=www --with-http_
    ssl_module --with-pcre && make && make install
RUN echo "daemon off;" >> /etc/nginx.conf
#EXPOSE映射端口
EXPOSE 80
#CMD运行以下命令
CMD ["nginx"]
```

15.2　Spring Cloud 核心组件整合

我们将模拟一个业务场景并将 Spring Cloud 中的核心组件进行整合。规划如下 8 个项目：

- registry——服务注册与发现；
- config——外部配置；

- monitor——监控；
- zipkin——分布式跟踪；
- gateway——代理所有微服务的接口网关；
- auth-service——OAuth2 认证服务；
- svca-service——业务服务 A；
- svcb-service——业务服务 B。

其中每个服务之间的调用关系如图 15-1 所示。

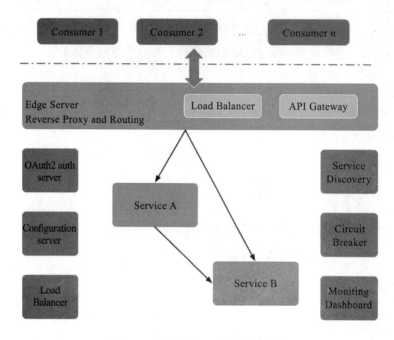

图 15-1　项目间相互调用依赖关系

其中，每个项目包含的技术点如图 15-2 所示。

1. Maven 父项目

首先新建父项目 spring-boot-cloud。由于通常 Maven 父项目中不需要含有业务代码，所以仅建立 pom.xml 即可。为了节省篇幅，仅列出 pom.xml 的关键点。

1）parent 的配置：配置 Maven 依赖的父坐标。

```xml
<parent>
    <groupId>org.springframework.boot</groupId>
    <artifactId>spring-boot-starter-parent</artifactId>
        <version>1.5.9.RELEASE</version>
    <relativePath/>
</parent>
```

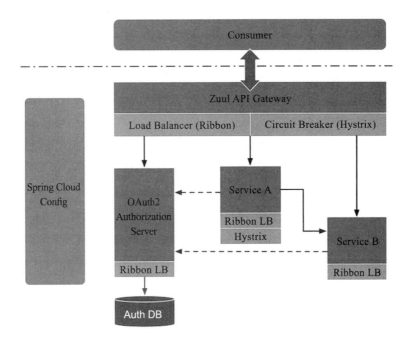

图 15-2　项目中包含的组件示意图

2）modules 的设置：配置所有的子项目列表。

```
<modules>
    <module>config</module>
    <module>registry</module>
    <module>gateway</module>
    <module>monitor</module>
    <module>auth-service</module>
    <module>svca-service</module>
    <module>svcb-service</module>
    <module>zipkin</module>
</modules>
```

3）dependencies 设置：设置用到的依赖。

```
<dependency>
        <groupId>org.springframework.boot</groupId>
        <artifactId>spring-boot-starter-actuator</artifactId>
    </dependency>
    <dependency>
        <groupId>org.springframework.boot</groupId>
        <artifactId>spring-boot-starter-test</artifactId>
        <scope>test</scope>
        <exclusions>
            <exclusion>
                <artifactId>asm</artifactId>
```

```xml
            <groupId>org.ow2.asm</groupId>
        </exclusion>
    </exclusions>
</dependency>
```

4）dependencyManagement 设置：依赖管理版本设置。

```xml
<dependencyManagement>
    <dependencies>
        <dependency>
            <groupId>de.codecentric</groupId>
            <artifactId>spring-boot-admin-server</artifactId>
            <version>${spring-boot-admin.version}</version>
        </dependency>
        <dependency>
            <groupId>de.codecentric</groupId>
            <artifactId>spring-boot-admin-server-ui</artifactId>
            <version>${spring-boot-admin.version}</version>
        </dependency>
        <dependency>
            <groupId>de.codecentric</groupId>
            <artifactId>spring-boot-admin-server-ui-hystrix</artifactId>
            <version>${spring-boot-admin.version}</version>
        </dependency>
        <dependency>
            <groupId>de.codecentric</groupId>
            <artifactId>spring-boot-admin-server-ui-turbine</artifactId>
            <version>${spring-boot-admin.version}</version>
        </dependency>
        <dependency>
            <groupId>de.codecentric</groupId>
            <artifactId>spring-boot-admin-server-ui-login</artifactId>
            <version>${spring-boot-admin.version}</version>
        </dependency>
        <dependency>
            <groupId>org.springframework.cloud</groupId>
            <artifactId>spring-cloud-dependencies</artifactId>
            <version>${spring.cloud.version}</version>
            <type>pom</type>
            <scope>import</scope>
        </dependency>
    </dependencies>
</dependencyManagement>
```

5）build 设置：打包参数的配置。

```xml
<build>
    <pluginManagement>
        <plugins>
            <plugin>
                <groupId>org.springframework.boot</groupId>
```

```xml
            <artifactId>spring-boot-maven-plugin</artifactId>
            <executions>
                <execution>
                    <goals>
                        <goal>repackage</goal>
                    </goals>
                </execution>
            </executions>
        </plugin>
        <plugin>
            <groupId>com.spotify</groupId>
            <artifactId>docker-maven-plugin</artifactId>
            <version>${docker.plugin.version}</version>
            <executions>
                <execution>
                    <phase>package</phase>
                    <goals>
                        <goal>build</goal>
                    </goals>
                </execution>
            </executions>
            <configuration>
                <imageName>${docker.image.prefix}/${project.artifactId}</imageName>
                <dockerDirectory>${project.basedir}/</dockerDirectory>
                <resources>
                    <resource>
                        <targetPath>/</targetPath>
                        <directory>${project.build.directory}</directory>
                        <include>${project.build.finalName}.jar</include>
                    </resource>
                </resources>
            </configuration>
        </plugin>
    </plugins>
    </pluginManagement>
</build>
```

2. 下游业务服务项目

我们模拟 Service B 为项目中的业务模块，处在业务下游将会被 Service A 调用。在 Service B 中我们就不写具体的业务实现了，以控制台打印来进行模拟业务行为。

创建 Service B 对应的 pom.xml，为了节省篇幅，仅列出依赖部分：

```xml
<dependencies>
    <dependency>
        <groupId>org.springframework.cloud</groupId>
        <artifactId>spring-cloud-starter-oauth2</artifactId>
    </dependency>

    <dependency>
```

```xml
        <groupId>org.springframework.cloud</groupId>
        <artifactId>spring-cloud-starter-eureka</artifactId>
</dependency>

<dependency>
        <groupId>org.springframework.cloud</groupId>
        <artifactId>spring-cloud-starter-sleuth</artifactId>
</dependency>
<dependency>
        <groupId>org.springframework.cloud</groupId>
        <artifactId>spring-cloud-sleuth-stream</artifactId>
</dependency>
<dependency>
        <groupId>org.springframework.cloud</groupId>
        <artifactId>spring-cloud-stream-binder-rabbit</artifactId>
</dependency>

<dependency>
        <groupId>org.springframework.cloud</groupId>
        <artifactId>spring-cloud-starter-config</artifactId>
</dependency>
<dependency>
        <groupId>org.springframework.cloud</groupId>
        <artifactId>spring-cloud-starter-bus-amqp</artifactId>
</dependency>
<dependency>
        <groupId>org.springframework.cloud</groupId>
        <artifactId>spring-cloud-starter-feign</artifactId>
</dependency>
<dependency>
        <groupId>org.springframework.cloud</groupId>
        <artifactId>spring-cloud-starter-hystrix</artifactId>
</dependency>
<dependency>
        <groupId>org.springframework.cloud</groupId>
        <artifactId>spring-cloud-netflix-hystrix-stream</artifactId>
</dependency>
<dependency>
        <groupId>org.springframework.cloud</groupId>
        <artifactId>spring-cloud-starter-ribbon</artifactId>
</dependency>
<dependency>
        <groupId>org.springframework.boot</groupId>
        <artifactId>spring-boot-starter-aop</artifactId>
</dependency>
<dependency>
        <groupId>org.springframework.retry</groupId>
        <artifactId>spring-retry</artifactId>
</dependency>
<dependency>
```

```xml
        <groupId>org.jolokia</groupId>
        <artifactId>jolokia-core</artifactId>
    </dependency>
</dependencies>
```

新建 main 函数及其类标注上需要的注解：

```java
@EnableDiscoveryClient
@EnableFeignClients
@SpringBootApplication
@EnableCircuitBreaker
@EnableOAuth2Client
public class ServiceBApplication {

    public static void main(String[] args) {
        SpringApplication.run(ServiceBApplication.class, args);
    }
}
```

实现一个 Controller，接受 HTTP 请求并打印输出 msg 信息：

```java
@RefreshScope
@RestController
public class ServiceBController {
    @Autowired
    EurekaDiscoveryClient discoveryClient;
    @Value("${msg:unknown}")
    private String msg;
    @GetMapping(value = "/")
    public String printServiceB() {
        ServiceInstance serviceInstance = discoveryClient.getLocalServiceInstance();
        return serviceInstance.getServiceId() + " (" + serviceInstance.getHost()
            + ":" + serviceInstance.getPort() + ")" + "===>Say " + msg;
    }
}
```

为了集成 OAuth2 的鉴权功能，实现一个配置类，并实现 Spring Security 下 ResourceServer ConfigurerAdapter 类的 configure 方法拦截所有请求进行验证：

```java
@Configuration
@EnableResourceServer
@EnableGlobalMethodSecurity(prePostEnabled = true, jsr250Enabled = true)
public class ServiceBConfiguration extends ResourceServerConfigurerAdapter {
    @Override
    public void configure(HttpSecurity http) throws Exception {
        http.authorizeRequests()
            .anyRequest().authenticated();
    }
}
```

新建 bootstrap.yml 进行配置 Config Server 地址等相关参数：

```yaml
spring:
    application:
        name: svcb-service
    cloud:
        config:
            uri: http://config:8888
            fail-fast: true
            username: user
            password: ${CONFIG_SERVER_PASSWORD:password}
            retry:
                initial-interval: 2000
                max-interval: 10000
                multiplier: 2
                max-attempts: 10
```

3. 上游业务工程

模拟 Service A 为业务模块中的上游链路模块,将会对 Service B 进行服务请求。首先创建 Service A 对应的 pom.xml,为了节省篇幅,仅列出依赖部分:

```xml
<dependencies>
    <dependency>
        <groupId>org.springframework.cloud</groupId>
        <artifactId>spring-cloud-starter-oauth2</artifactId>
    </dependency>

    <dependency>
        <groupId>org.springframework.cloud</groupId>
        <artifactId>spring-cloud-starter-eureka</artifactId>
    </dependency>

    <dependency>
        <groupId>org.springframework.cloud</groupId>
        <artifactId>spring-cloud-starter-sleuth</artifactId>
    </dependency>
    <dependency>
        <groupId>org.springframework.cloud</groupId>
        <artifactId>spring-cloud-sleuth-stream</artifactId>
    </dependency>
    <dependency>
        <groupId>org.springframework.cloud</groupId>
        <artifactId>spring-cloud-stream-binder-rabbit</artifactId>
    </dependency>

    <dependency>
        <groupId>org.springframework.cloud</groupId>
        <artifactId>spring-cloud-starter-config</artifactId>
    </dependency>
    <dependency>
        <groupId>org.springframework.cloud</groupId>
        <artifactId>spring-cloud-starter-bus-amqp</artifactId>
```

```xml
        </dependency>
        <dependency>
            <groupId>org.springframework.cloud</groupId>
            <artifactId>spring-cloud-starter-feign</artifactId>
        </dependency>
        <dependency>
            <groupId>org.springframework.cloud</groupId>
            <artifactId>spring-cloud-starter-hystrix</artifactId>
        </dependency>
        <dependency>
            <groupId>org.springframework.cloud</groupId>
            <artifactId>spring-cloud-netflix-hystrix-stream</artifactId>
        </dependency>
        <dependency>
            <groupId>org.springframework.cloud</groupId>
            <artifactId>spring-cloud-starter-ribbon</artifactId>
        </dependency>
        <dependency>
            <groupId>org.springframework.boot</groupId>
            <artifactId>spring-boot-starter-aop</artifactId>
        </dependency>
        <dependency>
            <groupId>org.springframework.retry</groupId>
            <artifactId>spring-retry</artifactId>
        </dependency>
        <dependency>
            <groupId>org.jolokia</groupId>
            <artifactId>jolokia-core</artifactId>
        </dependency>
</dependencies>
```

先实现 main 函数及其类：

```java
@EnableDiscoveryClient
@EnableFeignClients
@SpringBootApplication
@EnableCircuitBreaker
@EnableOAuth2Client
public class ServiceAApplication {

    public static void main(String[] args) {
        SpringApplication.run(ServiceAApplication.class, args);
    }

}
```

实现调用 Service B 的 FeignClient，并配置熔断的实现：

```java
@FeignClient(name = "svcb-service", fallback = ServiceBClient.ServiceBClientFallback.class)
    public interface ServiceBClient {
```

```
    @GetMapping(value = "/")
    String printServiceB();

    @Component
    class ServiceBClientFallback implements ServiceBClient {

        private static final Logger LOGGER = LoggerFactory.getLogger(ServiceBCli
            entFallback.class);

        @Override
        public String printServiceB() {
            LOGGER.info("异常发生，进入fallback方法");
            return "SERVICE B FAILED! - FALLING BACK";
        }
    }
}
```

与 Service B 一样，配置 SpringSecurity 的拦截范围，并配置 FeignClient 的 OAuth2 拦截器：

```
@Configuration
@EnableResourceServer
@EnableGlobalMethodSecurity(prePostEnabled = true, jsr250Enabled = true)
public class ServiceAConfiguration extends ResourceServerConfigurerAdapter {

    @Bean
    public RequestInterceptor oauth2FeignRequestInterceptor(OAuth2ClientContext
        oauth2ClientContext,
    ClientCredentialsResourceDetails resource) {
        return new OAuth2FeignRequestInterceptor(oauth2ClientContext, resource);
    }

    @Override
    public void configure(HttpSecurity http) throws Exception {
        http.authorizeRequests()
                .anyRequest().authenticated();
    }
}
```

最后实现 Service A 的 Controller，并在 API 实现中调用 Service B：

```
package cn.zhangxd.svca.controller;
@RefreshScope
@RestController
public class ServiceAController {

    @Value("${name:unknown}")
    private String name;

    @Autowired
```

```
    EurekaDiscoveryClient discoveryClient;
    @Autowired
    private ServiceBClient serviceBClient;

    @GetMapping(value = "/")
    public String printServiceA() {
        ServiceInstance serviceInstance = discoveryClient.getLocalServiceInstance();
        return serviceInstance.getServiceId() + " (" + serviceInstance.getHost()
            + ":" + serviceInstance.getPort() + ")" + "===>name:" + name +
            "<br/>" + serviceBClient.printServiceB();
    }

    @GetMapping(path = "/current")
    public Principal getCurrentAccount(Principal principal) {
        return principal;
    }
}
```

和 Service B 一样，编写 bootstrap.yml：

```
spring:
    application:
        name: svca-service
    cloud:
        config:
            uri: http://config:8888
            fail-fast: true
            username: user
            password: ${CONFIG_SERVER_PASSWORD:password}
            retry:
                initial-interval: 2000
                max-interval: 10000
                multiplier: 2
                max-attempts: 10
```

4. 注册中心工程

registry 工程为我们的服务注册中心工程，通过 Eureka Server 来实现。
先创建 pom 文件并配置 Eureka 相关依赖：

```xml
<dependencies>
    <dependency>
        <groupId>org.springframework.cloud</groupId>
        <artifactId>spring-cloud-starter-eureka-server</artifactId>
    </dependency>
    <dependency>
        <groupId>org.springframework.boot</groupId>
        <artifactId>spring-boot-starter-security</artifactId>
    </dependency>
</dependencies>
```

首先创建 Spring Boot 启动类，并标注上 Eureka Server 注解：

```java
@SpringBootApplication
@EnableEurekaServer
public class RegistryApplication {

    public static void main(String[] args) {
        SpringApplication.run(RegistryApplication.class, args);
    }
}
```

接着创建配置文件 application.yml，并指定初始化访问账号及密码：

```yaml
server:
    port: 8761

eureka:
    instance:
        hostname: registry
        prefer-ip-address: true
    client:
        registerWithEureka: false
        fetchRegistry: false
        service-url:
            defaultZone: http://${security.user.name}:${security.user.password}@${eureka.instance.hostname}:${server.port}/eureka/

security:
    user:
        name: user
        password: ${REGISTRY_SERVER_PASSWORD:password}
```

创建程序配置文件 bootstrap.yml：

```yaml
spring:
    application:
        name: registry
```

5. 配置中心工程

config 工程为我们的配置管理中心工程，通过 Spring Cloud Config 组件来实现。

首先创建 pom.xml 并配置 config 相关依赖：

```xml
<dependencies>
    <dependency>
        <groupId>org.springframework.cloud</groupId>
        <artifactId>spring-cloud-config-server</artifactId>
    </dependency>
    <dependency>
        <groupId>org.springframework.cloud</groupId>
        <artifactId>spring-cloud-config-monitor</artifactId>
```

```xml
    </dependency>
    <dependency>
        <groupId>org.springframework.cloud</groupId>
        <artifactId>spring-cloud-starter-stream-rabbit</artifactId>
    </dependency>
    <dependency>
        <groupId>org.springframework.cloud</groupId>
        <artifactId>spring-cloud-starter-eureka</artifactId>
    </dependency>
    <dependency>
        <groupId>org.springframework.boot</groupId>
        <artifactId>spring-boot-starter-security</artifactId>
    </dependency>
</dependencies>
```

创建 Config Server 主类：

```java
@EnableDiscoveryClient
@EnableConfigServer
@SpringBootApplication
public class ConfigApplication {

    public static void main(String[] args) {
        SpringApplication.run(ConfigApplication.class, args);
    }

}
```

创建配置文件 application.yml，并设置配置文件存储的路径：

```yaml
server:
    port: 8888

eureka:
    instance:
        hostname: registry
        prefer-ip-address: true
        metadata-map:
            user.name: ${security.user.name}
            user.password: ${security.user.password}
    client:
        service-url:
            defaultZone: http://user:${REGISTRY_SERVER_PASSWORD:password}@registry:8761/
                eureka/

spring:
    cloud:
        config:
            server:
                git:
                    uri: https://gitee.com/wawzw123/Spring CloudBookCode
```

```yaml
          search-paths: config-repo
  rabbitmq:
    host: rabbitmq

security:
  user:
    name: user
    password: ${CONFIG_SERVER_PASSWORD:password}
```

然后去 Git 中配置各个应用的配置文件。

创建通用配置文件 application.yml，配置各个组件相关参数如超时信息、鉴权服务器等：

```yaml
eureka:
  instance:
    hostname: registry
    prefer-ip-address: true
  client:
    service-url:
      defaultZone: http://user:${REGISTRY_SERVER_PASSWORD:password}@registry:8761/eureka/

hystrix:
  command:
    default:
      execution:
        isolation:
          thread:
            timeoutInMilliseconds: 10000

ribbon:
  ReadTimeout: 5000
  ConnectTimeout: 5000

spring:
  rabbitmq:
    host: rabbitmq
sleuth:
  sampler:
    percentage: 1
  integration:
    enabled: false
  scheduled:
    skip-pattern: "^org.*HystrixStreamTask$"

authserver:
  hostname: auth-service
  port: 5000
  contextPath: /uaa

security:
```

```yaml
      oauth2:
        resource:
          user-info-uri: http://${authserver.hostname}:${authserver.
            port}${authserver.contextPath}/current
```

配置鉴权应用相关信息 auth-service.yml 及默认使用的内存数据库：

```yaml
server:
    context-path: /uaa
    port: 5000

management:
    security:
        enabled: false
    context-path: /mgmt

eureka:
    instance:
        health-check-url-path: ${server.context-path}${management.context-
            path}/health
        status-page-url-path: ${server.context-path}${management.context-path}/info
        metadata-map:
            management.context-path: ${server.context-path}${management.context-path}
spring:
    datasource:
        url: jdbc:h2:mem:user
        driver-class-name: org.h2.Driver
    jpa:
        show-sql: true
```

配置 Zuul 应用的相关信息 gateway.yml 及路由转发策略：

```yaml
server:
    port: 8060

management:
    security:
        enabled: false
hystrix:
    command:
        default:
            execution:
                isolation:
                    thread:
                        timeoutInMilliseconds: 20000
ribbon:
    ReadTimeout: 10000
    ConnectTimeout: 10000

zuul:
    ignoredServices: '*'
```

```yaml
      routes:
        auth-service:
          path: /uaa/**
          stripPrefix: false
          sensitiveHeaders:
        svca-service:
          path: /svca/**
          sensitiveHeaders:
        svcb-service:
          path: /svcb/**
          sensitiveHeaders:
```

Service A 的配置文件为 svca-service.yml，配置认证服务器信息：

```yaml
server:
    port: 8080
  name: zhangxd
  eureka:
    instance:
      metadata-map:
        user.name: ${security.user.name}
        user.password: ${security.user.password}
  security:
    user:
      name: user
      password: password
    oauth2:
      client:
        clientId: svca-service
        clientSecret: ${security.user.password}
        accessTokenUri: http://${authserver.hostname}:${authserver.port}${authserver.contextPath}/oauth/token
        grant-type: client_credentials
        scope: server
```

Service B 的配置文件与 ServiceA 类似，是 svcb-service.yml：

```yaml
server:
    port: 8070
  msg: Hello
  eureka:
    instance:
      metadata-map:
        user.name: ${security.user.name}
        user.password: ${security.user.password}
  security:
    user:
      name: user
      password: password
    oauth2:
      client:
```

```
clientId: svcb-service
clientSecret: ${security.user.password}
accessTokenUri: http://${authserver.hostname}:${authserver.
    port}${authserver.contextPath}/oauth/token
grant-type: client_credentials
scope: server
```

6. 监控中心

接下来开发监控中心工程。我们会将几个组件的监控面板进行整合，达到一站式监控点效果。

先配置监控台需要的各个组件的依赖 pom.xml：

```
<dependencies>
    <dependency>
        <groupId>de.codecentric</groupId>
        <artifactId>spring-boot-admin-server</artifactId>
    </dependency>
    <dependency>
        <groupId>de.codecentric</groupId>
        <artifactId>spring-boot-admin-server-ui</artifactId>
    </dependency>
    <dependency>
        <groupId>de.codecentric</groupId>
        <artifactId>spring-boot-admin-server-ui-login</artifactId>
    </dependency>
    <dependency>
        <groupId>de.codecentric</groupId>
        <artifactId>spring-boot-admin-server-ui-hystrix</artifactId>
    </dependency>
    <dependency>
        <groupId>de.codecentric</groupId>
        <artifactId>spring-boot-admin-server-ui-turbine</artifactId>
    </dependency>
    <dependency>
        <groupId>org.springframework.cloud</groupId>
        <artifactId>spring-cloud-starter-turbine-stream</artifactId>
        <exclusions>
            <exclusion>
                <artifactId>netty-transport-native-epoll</artifactId>
                <groupId>io.netty</groupId>
            </exclusion>
            <exclusion>
                <artifactId>netty-codec-http</artifactId>
                <groupId>io.netty</groupId>
            </exclusion>
        </exclusions>
    </dependency>
    <dependency>
        <groupId>org.springframework.cloud</groupId>
```

```xml
        <artifactId>spring-cloud-starter-stream-rabbit</artifactId>
    </dependency>
    <dependency>
        <groupId>org.springframework.cloud</groupId>
        <artifactId>spring-cloud-starter-eureka</artifactId>
    </dependency>
    <dependency>
        <groupId>org.springframework.boot</groupId>
        <artifactId>spring-boot-starter-security</artifactId>
    </dependency>
</dependencies>
```

编写监控台集成的主类，并配置拦截 URL 的部分：

```java
@SpringBootApplication
@EnableDiscoveryClient
@EnableAdminServer
@EnableTurbineStream
public class MonitorApplication {

    public static void main(String[] args) {
        SpringApplication.run(MonitorApplication.class, args);
    }

    @Configuration
    public static class SecurityConfig extends WebSecurityConfigurerAdapter {
        @Override
        protected void configure(HttpSecurity http) throws Exception {
            http.formLogin().loginPage("/login.html").loginProcessingUrl("/
                login").permitAll();
            http.logout().logoutUrl("/logout");
            http.csrf().disable();
            http.authorizeRequests()
                .antMatchers("/login.html", "/**/*.css", "/img/**", "/third-party/**")
                .permitAll();
            http.authorizeRequests().antMatchers("/**").authenticated();

            http.httpBasic();
        }
    }
}
```

配置 monitor 的配置文件 application.yml，并配置控制台需要的参数：

```yaml
logging:
    level:
        org.springframework.cloud.netflix.zuul.filters.post.SendErrorFilter: error

server:
    port: 8040
```

```yaml
turbine:
    stream:
        port: 8041

eureka:
    instance:
        hostname: registry
        prefer-ip-address: true
        metadata-map:
            user.name: ${security.user.name}
            user.password: ${security.user.password}
    client:
        service-url:
            defaultZone: http://user:${REGISTRY_SERVER_PASSWORD:password}@registry:8761/
                eureka/

spring:
    rabbitmq:
        host: rabbitmq
    boot:
        admin:
            routes:
                endpoints: env,metrics,trace,dump,jolokia,info,configprops,trace,
                    logfile,refresh,flyway,liquibase,heapdump,loggers,auditevents,
                    hystrix.stream
            turbine:
                clusters: default
                location: http://monitor:${turbine.stream.port}

security:
    user:
        name: admin
        password: ${MONITOR_SERVER_PASSWORD:admin}
```

7. 日志追踪工程

接下来创建日志监控工程。先配置 pom.xml 并配置 Zipkin、Seluth 等依赖：

```xml
<dependencies>
    <dependency>
        <groupId>org.springframework.cloud</groupId>
        <artifactId>spring-cloud-sleuth-zipkin-stream</artifactId>
    </dependency>
    <dependency>
        <groupId>org.springframework.cloud</groupId>
        <artifactId>spring-cloud-starter-sleuth</artifactId>
    </dependency>
    <dependency>
        <groupId>org.springframework.cloud</groupId>
        <artifactId>spring-cloud-stream-binder-rabbit</artifactId>
    </dependency>
```

```xml
<dependency>
    <groupId>io.zipkin.java</groupId>
    <artifactId>zipkin-autoconfigure-ui</artifactId>
</dependency>
<dependency>
    <groupId>org.springframework.boot</groupId>
    <artifactId>spring-boot-starter-security</artifactId>
</dependency>
</dependencies>
```

编写主类,并标注上 Zipkin 相关的注解:

```java
@SpringBootApplication
@EnableZipkinStreamServer
public class ZipkinApplication {
    public static void main(String[] args) {
        SpringApplication.run(ZipkinApplication.class, args);
    }
}
```

配置 application.yml:

```yaml
spring:
  application:
    name: zipkin
  rabbitmq:
    host: rabbitmq
server:
  port: 9411
security:
  user:
    name: admin
    password: ${ZIPKIN_SERVER_PASSWORD:admin}
```

8. gateway

配置 Zuul 的相关依赖 pom.xml:

```xml
<dependencies>
    <dependency>
        <groupId>org.springframework.cloud</groupId>
        <artifactId>spring-cloud-starter-zuul</artifactId>
    </dependency>
    <dependency>
        <groupId>org.springframework.cloud</groupId>
        <artifactId>spring-cloud-starter-config</artifactId>
    </dependency>
    <dependency>
        <groupId>org.springframework.cloud</groupId>
        <artifactId>spring-cloud-starter-eureka</artifactId>
    </dependency>
    <dependency>
```

```xml
        <groupId>org.springframework.cloud</groupId>
        <artifactId>spring-cloud-starter-bus-amqp</artifactId>
    </dependency>
    <dependency>
        <groupId>org.springframework.cloud</groupId>
        <artifactId>spring-cloud-netflix-hystrix-stream</artifactId>
    </dependency>
    <dependency>
        <groupId>org.springframework.cloud</groupId>
        <artifactId>spring-cloud-starter-sleuth</artifactId>
    </dependency>
    <dependency>
        <groupId>org.springframework.cloud</groupId>
        <artifactId>spring-cloud-sleuth-stream</artifactId>
    </dependency>
    <dependency>
        <groupId>org.springframework.cloud</groupId>
        <artifactId>spring-cloud-stream-binder-rabbit</artifactId>
    </dependency>
    <dependency>
        <groupId>org.springframework.boot</groupId>
        <artifactId>spring-boot-starter-aop</artifactId>
    </dependency>
    <dependency>
        <groupId>org.springframework.retry</groupId>
        <artifactId>spring-retry</artifactId>
    </dependency>
    <dependency>
        <groupId>org.jolokia</groupId>
        <artifactId>jolokia-core</artifactId>
    </dependency>
</dependencies>
```

标注上 Zuul 的注解，编写主类：

```java
@SpringBootApplication
@EnableDiscoveryClient
@EnableZuulProxy
public class GatewayApplication {
    public static void main(String[] args) {
        SpringApplication.run(GatewayApplication.class, args);
    }
}
```

编写配置文件 application.yml：

```yaml
spring:
  application:
    name: gateway
  cloud:
    config:
```

```
            uri: http://config:8888
        fail-fast: true
        username: user
        password: ${CONFIG_SERVER_PASSWORD:password}
        retry:
            initial-interval: 2000
            max-interval: 10000
            multiplier: 2
            max-attempts: 10
```

9. 统一认证工程

我们将 OAuth 工程作为本项目中的认证工程，提供统一的鉴权服务，对各个服务之间的调用提供标准的 OAuth2 认证服务。

首先编写认证服务器的 pom.xml，添加相关依赖：

```xml
<dependencies>
    <dependency>
        <groupId>org.springframework.boot</groupId>
        <artifactId>spring-boot-starter-data-jpa</artifactId>
    </dependency>
    <dependency>
        <groupId>com.h2database</groupId>
        <artifactId>h2</artifactId>
        <scope>runtime</scope>
    </dependency>

    <dependency>
        <groupId>org.springframework.cloud</groupId>
        <artifactId>spring-cloud-starter-oauth2</artifactId>
    </dependency>

    <dependency>
        <groupId>org.springframework.cloud</groupId>
        <artifactId>spring-cloud-starter-eureka</artifactId>
    </dependency>

    <dependency>
        <groupId>org.springframework.cloud</groupId>
        <artifactId>spring-cloud-starter-sleuth</artifactId>
    </dependency>
    <dependency>
        <groupId>org.springframework.cloud</groupId>
        <artifactId>spring-cloud-sleuth-stream</artifactId>
    </dependency>
    <dependency>
        <groupId>org.springframework.cloud</groupId>
        <artifactId>spring-cloud-stream-binder-rabbit</artifactId>
    </dependency>
```

```xml
<dependency>
    <groupId>org.springframework.cloud</groupId>
    <artifactId>spring-cloud-starter-config</artifactId>
</dependency>
<dependency>
    <groupId>org.springframework.cloud</groupId>
    <artifactId>spring-cloud-starter-bus-amqp</artifactId>
</dependency>
<dependency>
    <groupId>org.springframework.cloud</groupId>
    <artifactId>spring-cloud-starter-hystrix</artifactId>
</dependency>
<dependency>
    <groupId>org.springframework.cloud</groupId>
    <artifactId>spring-cloud-netflix-hystrix-stream</artifactId>
</dependency>
<dependency>
    <groupId>org.springframework.cloud</groupId>
    <artifactId>spring-cloud-starter-ribbon</artifactId>
</dependency>
<dependency>
    <groupId>org.springframework.boot</groupId>
    <artifactId>spring-boot-starter-aop</artifactId>
</dependency>
<dependency>
    <groupId>org.springframework.retry</groupId>
    <artifactId>spring-retry</artifactId>
</dependency>
<dependency>
    <groupId>org.jolokia</groupId>
    <artifactId>jolokia-core</artifactId>
</dependency>
</dependencies>
```

编写 application.yml 配置文件:

```yaml
spring:
    application:
        name: auth-service
    cloud:
        config:
            uri: http://config:8888
            fail-fast: true
            username: user
            password: ${CONFIG_SERVER_PASSWORD:password}
            retry:
                initial-interval: 2000
                max-interval: 10000
                multiplier: 2
                max-attempts: 10
```

编写主程序：

```java
@SpringBootApplication
@EnableDiscoveryClient
public class AuthApplication {

    public static void main(String[] args) {
        SpringApplication.run(AuthApplication.class, args);
    }
}
```

编写 OAuth 的配置类：

```java
@Configuration
@EnableAuthorizationServer
public class OAuthConfiguration extends AuthorizationServerConfigurerAdapter {

    @Autowired
    private AuthenticationManager auth;

    @Autowired
    private DataSource dataSource;

    private BCryptPasswordEncoder passwordEncoder = new BCryptPasswordEncoder();

    @Bean
    public JdbcTokenStore tokenStore() {
        return new JdbcTokenStore(dataSource);
    }

    @Override
    public void configure(AuthorizationServerSecurityConfigurer security)
            throws Exception {
        security.passwordEncoder(passwordEncoder);
    }

    @Override
    public void configure(AuthorizationServerEndpointsConfigurer endpoints)
            throws Exception {
        endpoints
                .authenticationManager(auth)
                .tokenStore(tokenStore())
        ;
    }

    @Override
    public void configure(ClientDetailsServiceConfigurer clients)
            throws Exception {

        clients.jdbc(dataSource)
                .passwordEncoder(passwordEncoder)
```

```java
                    .withClient("client")
                    .secret("secret")
                    .authorizedGrantTypes("password", "refresh_token")
                    .scopes("read", "write")
                    .accessTokenValiditySeconds(3600) // 1 hour
                    .refreshTokenValiditySeconds(2592000) // 30 days
                    .and()
                    .withClient("svca-service")
                    .secret("password")
                    .authorizedGrantTypes("client_credentials", "refresh_token")
                    .scopes("server")
                    .and()
                    .withClient("svcb-service")
                    .secret("password")
                    .authorizedGrantTypes("client_credentials", "refresh_token")
                    .scopes("server")
            ;

    }

    @Configuration
    @Order(-20)
    protected static class AuthenticationManagerConfiguration extends GlobalAuthenticationConfigurerAdapter {

        @Autowired
        private DataSource dataSource;

        @Override
        public void init(AuthenticationManagerBuilder auth) throws Exception {
            auth.jdbcAuthentication().dataSource(dataSource)
                    .withUser("dave").password("secret").roles("USER")
                    .and()
                    .withUser("anil").password("password").roles("ADMIN")
            ;
        }
    }
}
```

编写拦截配置类：

```java
@Configuration
@EnableResourceServer
public class ResourceServerConfiguration extends ResourceServerConfigurerAdapter {

    @Override
    public void configure(HttpSecurity http) throws Exception {
        http
                .requestMatchers().antMatchers("/current")
                .and()
                .authorizeRequests()
```

```
            .antMatchers("/current").access("#oauth2.hasScope('read')");
    }
}
```

最后暴露 Controller：

```
@RestController
@RequestMapping("/")
public class UserController {

    @GetMapping(value = "/current")
    public Principal getUser(Principal principal) {
        return principal;
    }
}
```

15.3　Dockerfile 编写

基于 15.1.2 节所述，应该可以看懂并编写常用的 Dockerfile。为了让项目整体能够通过 Docker Compose 一键启动，接下来针对每一个应用一步步地编写独立的 Dockerfile，配置端口、启动命令等相关参数，然后配置 Docker Compose 的支持。

统一将 java:8 镜像作为基础镜像，并在其之上开始编写。

1）auth-service 应用对应 Dockerfile：

```
#指定以"Java:8"为基础镜像
FROM java:8
#挂载tmp目录
VOLUME /tmp
#将我们打包好的应用的Jar文件复制进容器内
ADD ./target/auth-service.jar /app.jar
#将Jar文件的时间标签更新为系统当前的时间
RUN bash -c 'touch /app.jar'
#指定暴露端口5000
EXPOSE 5000
#容器启动后,执行命令启动应用Jar
ENTRYPOINT ["java","-jar","/app.jar"]
```

2）gateway 应用对应 Dockerfile（与上文应用结构相同，不再添加注释，参考上文）：

```
FROM java:8
VOLUME /tmp
ADD ./target/gateway.jar /app.jar
RUN bash -c 'touch /app.jar'
EXPOSE 8060
ENTRYPOINT ["java","-jar","/app.jar"]
```

3）monitor 应用对应 Dockerfile（与上文应用结构相同，不再添加注释，参考上文）：

```
FROM java:8
VOLUME /tmp
ADD ./target/monitor.jar /app.jar
RUN bash -c 'touch /app.jar'
EXPOSE 8040 8041
ENTRYPOINT ["java","-jar","/app.jar"]
```

4) registry 应用对应 Dockerfile（与上文应用结构相同，不再添加注释，参考上文）：

```
FROM java:8
VOLUME /tmp
ADD ./target/registry.jar /app.jar
RUN bash -c 'touch /app.jar'
EXPOSE 8761
ENTRYPOINT ["java","-jar","/app.jar"]
```

5) svca-service 应用对应 Dockerfile（与上文应用结构相同，不再添加注释，参考上文）：

```
FROM java:8
VOLUME /tmp
ADD ./target/svca-service.jar /app.jar
RUN bash -c 'touch /app.jar'
EXPOSE 8080
ENTRYPOINT ["java","-jar","/app.jar"]
```

6) svcb-service 应用对应 Dockerfile（与上文应用结构相同，不再添加注释，参考上文）：

```
FROM java:8
VOLUME /tmp
ADD ./target/svcb-service.jar /app.jar
RUN bash -c 'touch /app.jar'
EXPOSE 8070
ENTRYPOINT ["java","-jar","/app.jar"]
```

7) zipkin 应用对应 Dockerfile（与上文应用结构相同，不再添加注释，参考上文）：

```
FROM java:8
VOLUME /tmp
ADD ./target/zipkin.jar /app.jar
RUN bash -c 'touch /app.jar'
EXPOSE 9411
ENTRYPOINT ["java","-jar","/app.jar"]
```

8) config 应用对应 Dockerfile（与上文应用结构相同，不再添加注释，参考上文）：

```
FROM java:8
VOLUME /tmp
ADD ./target/config.jar /app.jar
RUN bash -c 'touch /app.jar'
EXPOSE 8888
ENTRYPOINT ["java","-jar","/app.jar"]
```

9）Docker Compose 文件编写设置启动顺序并针对每个应用分别设置端口号：

```
version: '2'
services:
    rabbitmq:
        image: rabbitmq:3-management
        restart: always
        ports:
            - 15673:15672
    registry:
        image: spring-boot-cloud/registry
        ports:
            - "8761:8761"
    config:
        image: spring-boot-cloud/config
        ports:
            - "8888:8888"
    monitor:
        image: spring-boot-cloud/monitor
        ports:
            - "8040:8040"
    zipkin:
        image: spring-boot-cloud/zipkin
        ports:
            - "9411:9411"
    gateway:
        image: spring-boot-cloud/gateway
        ports:
            - "8060:8060"
    auth-service:
        image: spring-boot-cloud/auth-service
        ports:
            - "5000:5000"
    svca-service:
        image: spring-boot-cloud/svca-service
    svcb-service:
        image: spring-boot-cloud/svcb-service
```

15.4 启动与接口测试

整体项目的所有代码及配置文件等已经全部开发完毕，就可以启动整个微服务项目。不像单体应用，分布式项目的启动步骤比较多，且相互之间是有依赖性的。

1. 启动项目

（1）使用 Docker 快速启动

步骤 1 配置 Docker 环境。

步骤 2 执行 Maven 命令 "mvn clean package" 打包项目及 Docker 镜像。
步骤 3 在项目根目录下执行 docker-compose up -d，启动所有项目。

（2）本地手动启动

步骤 1 配置 RabbitMQ。

步骤 2 修改 hosts 将主机名指向本地：127.0.0.1 registry config monitor rabbitmq auth-service，或者修改各服务配置文件中的相应主机名为本地 IP。

步骤 3 启动 registry、config、monitor、zipkin。

步骤 4 启动 gateway、auth-service、svca-service、svcb-service。

2. 接口测试

在项目启动完成之后，通过 curl 工具发起 HTTP 请求，来模拟一个从认证开始直到获取到真实业务数据的流程。

步骤 1 获取 Token。

```
curl -X POST -vu client:secret http://localhost:8060/uaa/oauth/token -H
    "Accept: application/json" -d "password=password&username=anil&grant_
    type=password&scope=read%20write"
```

步骤 2 返回如下格式数据。

```
{
    "access_token": "eac56504-c4f0-4706-b72e-3dc3acdf45e9",
    "token_type": "bearer",
    "refresh_token": "dd1007dc-603c-4309-965d-370b15aa1aeb",
    "expires_in": 3599,
    "scope": "read write"
}
```

步骤 3 使用 access token 访问 Service A 接口。

```
curl -i -H "Authorization: Bearer eac56504-c4f0-4706-b72e-3dc3acdf45e9" http://
    localhost:8060/svca
```

步骤 4 返回如下数据。

```
svca-service (172.18.0.8:8080)===>name:zhangxd
svcb-service (172.18.0.2:8070)===>Say Hello
```

步骤 5 使用 access token 访问 Service B 接口。

```
curl -i -H "Authorization: Bearer eac56504-c4f0-4706-b72e-3dc3acdf45e9" http://
    localhost:8060/svcb
```

步骤 6 返回如下数据。

```
svcb-service (172.18.0.2:8070)===>Say Hello
```

步骤 7 使用 refresh token 刷新 Token。

```
curl -X POST -vu client:secret http://localhost:8060/uaa/oauth/token -H "Accept:
    application/json" -d "grant_type=refresh_token&refresh_token=da1007dc-683c-
    4309-965d-370b15aa4aeb"
```

步骤 8 返回更新后的 Token。

```
{
    "access_token": "63ff57ce-f140-482e-ba7e-b6f29df35c88",
    "token_type": "bearer",
    "refresh_token": "da1007dc-683c-4309-965d-370b15aa4aeb",
    "expires_in": 3599,
    "scope": "read write"
}
```

步骤 9 刷新配置。

```
curl -X POST -vu user:password http://localhost:8888/bus/refresh
```

最后，可以登录如调用链监控、注册中心控制台等界面，查看执行效果。

15.5 小结

本章通过一个简单的项目案例将 Spring Cloud 核心组件串联起来，并整合了服务鉴权功能，将其以 Docker 化的形式进行了封装打包。希望本章使读者能够在实战中得出对 Spring Cloud 更深入、更实际的理解，对没有接触过 Docker 的读者进行引导。

后　　记

　　Spring Cloud 体系由 Spring 在 2015 年年初推出，至今已经三年多，其子项目在不断增加与完善，其所包含的功能点已经基本满足整个微服务体系的需求。加上 Spring 本身强大的技术后盾和活跃的技术社区氛围，Spring Cloud 已经形成了一股势不可挡的力量，迅速扩张、覆盖至整个互联网技术公司。尝试调研使用并在生产环境落地 Spring Cloud 的互联网公司越来越多，Netflix 公司也在不断迭代与完善其组件，源源不断地供给 Spring Cloud 更新的源泉。

　　近年来微服务的概念就像当年的 SOA，如风口上的"猪"。这不只是一阵概念的热度。在需求方，如互联网公司、软件公司等随着自身的业务体系越来越大，对技术层面的迭代效率要求也越来越高，但由于业务体系大，系统复杂度越来越高，导致我们不得不将服务越差越细，但是同时带来的分布式场景下的衍射问题暴露得越来越多。我们迫切地需要一整套完善、健全、稳定的分布式服务框架来为我们解决这些问题。Docker 的出现让运维部署层面看到了希望。然而在具体的开发层面，Spring Cloud 问世之前各大互联网公司都有自己的服务框架，大多数是在参考业内的基础上自主研发，导致各个公司重复造轮子，浪费了资源并降低了效率，一旦开发人员更替有很大可能停止维护又重新造个轮子。技术人员在更换工作的时候要重新熟悉新公司的整个技术体系，带来的成本也很高，像很多年前 SSH 一统江湖的时候每个公司都是 SSH，在技术层面形成了大一统之势。利用 Spring Cloud，无技术体系切换的学习成本，由 Apache、Spring 等开源组织维护技术框架，互联网公司可以更多专注于自身公司的业务层面。

　　我们迫切地希望 Spring 与各大技术公司能够形成一个良性的循环，使用者越多，技术社区越活跃，Spring Cloud 的更新迭代速度越快，功能越完善、越稳定、越成熟，然后会带来更多的使用者。

Appendix 附录

配置汇总

名称	默认	描述
encrypt.fail-on-error	true	如果存在加密或解密错误，进程将失败
encrypt.key		对称密钥。生产环境建议使用密钥库
encrypt.key-store.alias		密码库中的钥匙别名
encrypt.key-store.location		密钥存储文件的位置，例如 classpath:/keystore.jks
encrypt.key-store.password		锁定密钥库的密码
encrypt.key-store.secret		秘密保护密钥（默认为密码相同）
encrypt.rsa.algorithm		使用RSA算法（DEFAULT或OEAP）。一旦设置不改变它（或现有的密码将不可解密）
encrypt.rsa.salt	deadbeef	Salt用于加密密文的校验用的随机密码
encrypt.rsa.strong	false	标识是否应该在内部使用"强"AES加密。如果为真，则将GCM算法应用于AES加密字节。默认值为false（在这种情况下使用"标准"CBC代替）
endpoints.bus.enabled		
endpoints.bus.id		
endpoints.bus.sensitive		
endpoints.consul.enabled		
endpoints.consul.id		
endpoints.consul.sensitive		

（续）

名　称	默　认	描　述
endpoints.env.post.enabled	true	允许通过 POST 请求将环境更改为 /env
endpoints.features.enabled		
endpoints.features.id		
endpoints.features.sensitive		
endpoints.pause.enabled	true	启用 / 暂停端点（发送 Lifecycle.stop()）
endpoints.pause.id		
endpoints.pause.sensitive		
endpoints.refresh.enabled	true	启用 / refresh 端点刷新配置并重新初始化刷新作用域 bean
endpoints.refresh.id		
endpoints.refresh.sensitive		
endpoints.restart.enabled	true	启用 / restart 端点重新启动应用程序上下文
endpoints.restart.id		
endpoints.restart.pause-endpoint.enabled		
endpoints.restart.pause-endpoint.id		
endpoints.restart.pause-endpoint.sensitive		
endpoints.restart.resume-endpoint.enabled		
endpoints.restart.resume-endpoint.id		
endpoints.restart.resume-endpoint.sensitive		
endpoints.restart.sensitive		
endpoints.restart.timeout	0	
endpoints.resume.enabled	true	启用 / resume 端点（发送 Lifecycle.start()）
endpoints.resume.id		
endpoints.resume.sensitive		
endpoints.zookeeper.enabled	true	启用 / zookeeper 端点来检查 ZooKeeper 的状态
eureka.client.allow-redirects	false	指示服务器是否可以将客户端请求重定向。如果设置为 false，服务器将直接处理请求；如果设置为 true，则可能会向客户端发送 HTTP 重定向
eureka.client.availability-zones		获取此实例所在区域的可用性区域列表（用于 AWS 数据中心）
eureka.client.backup-registry-impl		获取执行 BackupRegistry 的实现的名称，以便在 Eureka 客户端启动时将此注册信息作为调用方的备份

(续)

名称	默认	描述
eureka.client.cache-refresh-executor-exponential-back-off-bound	10	缓存刷新执行器相关属性。在发生超时的情况下，表示重试延迟的最大乘数值
eureka.client.cache-refresh-executor-thread-pool-size	2	cacheRefreshExecutor 初始化的线程池大小
eureka.client.client-data-accept		客户端数据接收时 EurekaAccept 的名称
eureka.client.decoder-name		配置解码器名称
eureka.client.disable-delta	false	指示 Eureka 客户端是否禁用增量提取
eureka.client.dollar-replacement	_	在 Eureka 服务器的序列化/反序列化信息期间，获取 $ 符号的替换字符串
eureka.client.enabled	true	标记以指示启用 Eureka 客户端
eureka.client.encoder-name		配置编码器名称
eureka.client.escape-char-replacement	__	在 Eureka 服务器的序列化/反序列化信息期间获取下划线符号 _ 的替换字符串
eureka.client.eureka-connection-idle-timeout-seconds	30	表示到 Eureka 服务器的 HTTP 连接可以在关闭之前保持空闲状态的时间（以 s 为单位）
eureka.client.eureka-server-connect-timeout-seconds	5	指在连接到 Eureka 服务器需要超时之前等待（以秒为单位）的时间。请注意，客户端中的连接由 org.apache.http.client.HttpClient 汇总，此设置会影响实际的连接创建以及从池中获取连接的等待时间
eureka.client.eureka-server-d-n-s-name		获取要查询的 DNS 名称来获得 Eureka 服务器，此配置只有在 Eureka 服务器 IP 地址列表是在 DNS 中才会用到
eureka.client.eureka-server-port		获取服务 url 的端口
eureka.client.eureka-server-read-timeout-seconds	8	指示从 Eureka 服务器读取之前需要等待（s）多久才能超时
eureka.client.eureka-server-total-connections	200	获取从 Eureka 客户端到所有 Eureka 服务器允许的总连接数
eureka.client.eureka-server-total-connections-per-host	50	获取从 Eureka 客户端到 Eureka 服务器主机允许的总连接数
eureka.client.eureka-server-u-r-l-context		表示 Eureka 注册中心的路径，如果配置为 Eureka，则为 http://ip:port/eureka/，在 Eureka 的配置文件中加入此配置表示 Eureka 作为客户端向注册中心注册，从而构成 Eureka 集群。此配置只有在 Eureka 服务器 IP 地址列表是在 DNS 中才会用到，默认为 null
eureka.client.eureka-service-url-poll-interval-seconds	0	表示轮询对 Eureka 服务器信息进行刷新的频率（以 s 为单位）

（续）

名称	默认	描述
eureka.client.fetch-registry	true	指该客户端是否应从 Eureka 服务器获取 Eureka 注册信息
eureka.client.fetch-remote-regions-registry		获取 Eureka 注册信息的区域列表
eureka.client.filter-only-up-instances	true	是否在仅具有 InstanceStatus UP 状态的实例的过滤应用程序之后获取应用程序
eureka.client.g-zip-content	true	是否压缩从 Eureka 服务器获取的内容
eureka.client.heartbeat-executor-exponential-back-off-bound	10	心跳超时重试延迟时间的最大乘数值，默认 10
eureka.client.heartbeat-executor-thread-pool-size	2	心跳线程池初始化的大小
eureka.client.initial-instance-info-replication-interval-seconds	40	初始化实例信息同步到 Eureka 服务器的间隔时间（以 s 为单位）
eureka.client.instance-info-replication-interval-seconds	30	将实例信息同步到 Eureka 服务器的间隔时间（以 s 为单位）
eureka.client.log-delta-diff	false	是否记录 Eureka 服务器和客户端之间在注册表信息方面的差异
eureka.client.on-demand-update-status-change	true	如果设置为 true，则通过 ApplicationInfo-Manager 将本地状态更新
eureka.client.prefer-same-zone-eureka	true	实例是否使用同一 Zone 里的 Eureka 服务器
eureka.client.property-resolver		属性解析器
eureka.client.proxy-host		获取代理主机
eureka.client.proxy-password		获取代理密码
eureka.client.proxy-port		获取代理端口
eureka.client.proxy-user-name		获取代理用户名
eureka.client.region	us-east-1	获取此实例所在的区域（用于 AWS 数据中心）
eureka.client.register-with-eureka	true	指示此实例是否应将其信息注册到 Eureka 服务器
eureka.client.registry-fetch-interval-seconds	30	指从 Eureka 服务器获取注册信息的频率（以 s 为单位）
eureka.client.registry-refresh-single-vip-address		指客户端是否只对单个 VIP 的注册信息感兴趣
eureka.client.service-url		配置 Eureka 地址
eureka.client.transport		
eureka.client.use-dns-for-fetching-service-urls	false	指示 Eureka 客户端是否使用 DNS 机制来获取要与之通信的 Eureka 服务器列表
eureka.dashboard.enabled	true	是否启用 Eureka 仪表板。默认值为 true

(续)

名 称	默 认	描 述
eureka.dashboard.path	/	Eureka 仪表板的路径。默认为"/"
eureka.instance.a-s-g-name		注册到注册中心的应用所属分组名称（AWS 服务器）
eureka.instance.app-group-name		获取要在 Eureka 中注册的应用组的名称
eureka.instance.appname	unknown	获取要在 Eureka 注册的应用的名称
eureka.instance.data-center-info		返回此实例部署的数据中心。如果实例部署在 AWS 中，则此信息用于获取一些 AWS 特定实例信息
eureka.instance.default-address-resolution-order	[]	默认地址解析顺序
eureka.instance.environment		该服务实例环境配置
eureka.instance.health-check-url		获取此实例的健康检查绝对路径 URL
eureka.instance.health-check-url-path	/health	获取此实例的健康检查相对路径 URL
eureka.instance.home-page-url		获取此实例的主页绝对路径 URL
eureka.instance.home-page-url-path	/	获取此实例的主页相对路径 URL
eureka.instance.host-info		
eureka.instance.hostname		配置主机名
eureka.instance.inet-utils		
eureka.instance.initial-status		该服务实例注册到 Eureka Server 的初始状态
eureka.instance.instance-enabled-onit	false	是否应在 Eureka 注册后立即启用实例以获取流量
eureka.instance.instance-id		获取 Eureka 注册的此实例的唯一 ID
eureka.instance.ip-address		获取实例的 IP
eureka.instance.lease-expiration-duration-in-seconds	90	Eureka 服务器在删除此实例之前收到最后一次心跳之后等待的时间
eureka.instance.lease-renewal-interval-in-seconds	30	该服务实例向注册中心发送心跳间隔（s）
eureka.instance.metadata-map		获取与此实例关联的元数据名称/值对
eureka.instance.namespace	eureka	获取命名空间
eureka.instance.non-secure-port	80	获取实例的非安全端口
eureka.instance.non-secure-port-enabled	true	是否启用非安全端口
eureka.instance.prefer-ip-address	false	相较于 hostname，是否优先使用服务实例的 IP 地址
eureka.instance.registry.default-open-for-traffic-count	1	Eureka Server 端属性，默认开启通信的数量

(续)

名称	默认	描述
eureka.instance.registry.expected-number-of-renews-per-min	1	Eureka Server 端属性，每分钟续约次数
eureka.instance.secure-health-check-url		获取健康检查 URL 的 HTTPS 地址
eureka.instance.secure-port	443	获取 HTTPS 端口
eureka.instance.secure-port-enabled	false	HTTPS 端口是否启用
eureka.instance.secure-virtual-host-name	unknown	HTTPS 的虚拟主机名
eureka.instance.status-page-url		该服务实例的状态检查地址（url），绝对地址
eureka.instance.status-page-url-path	/info	该服务实例的状态检查地址（url），相对地址
eureka.instance.virtual-host-name	unknown	实例的虚拟主机名
eureka.server.a-s-g-cache-expiry-timeout-ms	0	
eureka.server.a-s-g-query-timeout-ms	300	
eureka.server.a-s-g-update-interval-ms	0	
eureka.server.a-w-s-access-id		
eureka.server.a-w-s-secret-key		
eureka.server.batch-replication	false	节点之间数据复制是否采用批处理
eureka.server.binding-strategy		
eureka.server.delta-retention-timer-interval-in-ms	0	清理无效增量信息的时间间隔（ms）
eureka.server.disable-delta	false	禁用增量获取服务实例信息
eureka.server.disable-delta-for-remote-regions	false	
eureka.server.disable-transparent-fallback-to-other-region	false	
eureka.server.e-i-p-bind-rebind-retries	3	
eureka.server.e-i-p-binding-retry-interval-ms	0	
eureka.server.e-i-p-binding-retry-interval-ms-when-unbound	0	
eureka.server.enable-replicated-request-compression	false	复制数据请求时，数据是否压缩
eureka.server.enable-self-preservation	true	启用自我保护机制，默认为 true
eureka.server.eviction-interval-timer-in-ms	0	清除无效服务实例的时间间隔（ms）
eureka.server.g-zip-content-from-remote-region	true	当服务端支持压缩的情况下，是否支持将从服务端获取的信息进行压缩
eureka.server.json-codec-name		JSON 编解码器名称

(续)

名称	默认	描述
eureka.server.list-auto-scaling-groups-role-name	ListAutoScalingGroups	
eureka.server.log-identity-headers	true	是否记录登录日志
eureka.server.max-elements-in-peer-replication-pool	10000	备份池最大备份事件数量
eureka.server.max-elements-in-status-replication-pool	10000	状态备份池最大备份事件数量
eureka.server.max-idle-thread-age-in-minutes-for-peer-replication	15	节点之间信息同步线程最大空闲时间
eureka.server.max-idle-thread-in-minutes-age-for-status-replication	10	节点之间状态同步线程最大空闲时间
eureka.server.max-threads-for-peer-replication	20	节点之间信息同步最大线程数量
eureka.server.max-threads-for-status-replication	1	节点之间状态同步最大线程数量
eureka.server.max-time-for-replication	30000	节点之间信息复制最大通信时长（ms）
eureka.server.min-threads-for-peer-replication	5	节点之间信息复制最小线程数量
eureka.server.min-threads-for-status-replication	1	节点之间信息状态同步最小线程数量
eureka.server.number-of-replication-retries	5	节点之间数据复制时，可重试次数
eureka.server.peer-eureka-nodes-update-interval-ms	0	节点更新数据间隔时长
eureka.server.peer-eureka-status-refresh-time-interval-ms	0	节点之间状态刷新间隔时长
eureka.server.peer-node-connect-timeout-ms	200	节点之间连接超时时长
eureka.server.peer-node-connection-idle-timeout-seconds	30	节点之间连接后，空闲时长
eureka.server.peer-node-read-timeout-ms	200	节点之间数据读取超时时间
eureka.server.peer-node-total-connections	1000	集群中节点连接总数
eureka.server.peer-node-total-connections-per-host	500	节点之间连接，单机最大连接数量
eureka.server.prime-aws-replica-connections	true	
eureka.server.property-resolver		属性解析器名称
eureka.server.rate-limiter-burst-size	10	限流大小
eureka.server.rate-limiter-enabled	false	限流开关
eureka.server.rate-limiter-full-fetch-average-rate	100	平均请求速率
eureka.server.rate-limiter-privileged-clients		信任的客户端列表

(续)

名称	默认	描述
eureka.server.rate-limiter-registry-fetch-average-rate	500	服务注册与拉取的平均速率
eureka.server.rate-limiter-throttle-standard-clients	false	是否对标准客户端进行限流
eureka.server.registry-sync-retries	0	节点启动时，尝试获取注册信息的次数
eureka.server.registry-sync-retry-wait-ms	0	节点启动时，尝试获取注册信息的间隔时长
eureka.server.remote-region-app-whitelist		
eureka.server.remote-region-connect-timeout-ms	1000	
eureka.server.remote-region-connection-idle-timeout-seconds	30	
eureka.server.remote-region-fetch-thread-pool-size	20	
eureka.server.remote-region-read-timeout-ms	1000	
eureka.server.remote-region-registry-fetch-interval	30	
eureka.server.remote-region-total-connections	1000	
eureka.server.remote-region-total-connections-per-host	500	
eureka.server.remote-region-trust-store		
eureka.server.remote-region-trust-store-password	changeit	
eureka.server.remote-region-urls		
eureka.server.remote-region-urls-with-name		
eureka.server.renewal-percent-threshold	0.85	15分钟内续约服务的比例小于0.85，则开启自我保护机制，在此期间不会清除已注册的任何服务（即便是无效服务）
eureka.server.renewal-threshold-update-interval-ms	0	
eureka.server.response-cache-auto-expiration-in-seconds	80	注册信息缓存有效时长
eureka.server.response-cache-update-interval- ms	0	注册信息缓存更新间隔
eureka.server.retention-time-in-m-s-in-delta-queue	0	保留增量信息时长
eureka.server.route53-bind-rebind-retries	3	

(续)

名称	默认	描述
eureka.server.route53-binding-retry-interval-ms	0	
eureka.server.route53-domain-t-t-l	30	
eureka.server.sync-when-timestamp-differs	true	当时间戳不一致时，是否进行同步
eureka.server.use-read-only-response-cache	true	是否使用只读缓存策略
eureka.server.wait-time-in-ms-when-sync-empty	0	在 Eureka 服务器获取不到集群里对等服务器上的实例时，需要等待的时间
eureka.server.xml-codec-name		XML 编解码器名称
feign.compression.request.mime-types	[text/xml, application/xml, application/json]	支持的 MIME 类型列表
feign.compression.request.min-request-size	2048	请求内容压缩至最小内容大小
health.config.enabled	false	健康配置是否启用
health.config.time-to-live	0	默认缓存有效时间
hystrix.metrics.enabled	true	是否启用 Hystrix 指标
hystrix.metrics.polling-interval-ms	2000	后续轮询时间间隔
management.health.refresh.enabled	true	健康检查是否自动刷新
management.health.zookeeper.enabled	true	启用 ZooKeeper 的健康端点
netflix.atlas.batch-size	10000	
netflix.atlas.enabled	true	
netflix.atlas.uri		
netflix.metrics.servo.cache-warning-threshold	1000	当 ServoMonitorCache 达到这个阈值则触发警告
netflix.metrics.servo.registry-class	com.netflix.servo.BasicMonitorRegistry	Servo 监控的默认注册类
proxy.auth.load-balanced		
proxy.auth.routes		获取授权的路由策略组
spring.cloud.bus.ack.destination-service		指定 Bus-ack 的目标服务
spring.cloud.bus.ack.enabled	true	是否启用 ack 机制
spring.cloud.bus.destination	Spring CloudBus	指定 Bus 的目标 Topic 名称
spring.cloud.bus.enabled	true	是否启用
spring.cloud.bus.env.enabled	true	环境变化事件是否启用
spring.cloud.bus.refresh.enabled	true	关闭刷新事件是否启用
spring.cloud.bus.trace.enabled	false	打开 acks 跟踪的标志

(续)

名　称	默　认	描　述
spring.cloud.cloudfoundry.discovery.enabled	true	
spring.cloud.cloudfoundry.discovery.heartbeat-frequency	5000	
spring.cloud.cloudfoundry.discovery.org		
spring.cloud.cloudfoundry.discovery.password		
spring.cloud.cloudfoundry.discovery.space		
spring.cloud.cloudfoundry.discovery.url	https://api.run.pivotal.io	
spring.cloud.cloudfoundry.discovery.username		
spring.cloud.config.allow-override	true	是否允许重写
spring.cloud.config.authorization		是否启用鉴权模式
spring.cloud.config.discovery.enabled	false	是否启用服务发现功能
spring.cloud.config.discovery.service-id	configserver	注册 Eureka 时的 ID
spring.cloud.config.enabled	true	是否启用 Config
spring.cloud.config.fail-fast	false	是否启用快速失败机制
spring.cloud.config.label		用于获取远程配置的标签名称
spring.cloud.config.name		用于获取远程配置的应用名称
spring.cloud.config.override-none	false	配置是否允许覆盖
spring.cloud.config.override-system-properties	true	是否覆盖系统属性
spring.cloud.config.password		
spring.cloud.config.profile	default	
spring.cloud.config.retry.initial-interval	1000	初始重试间隔（以 ms 为单位）
spring.cloud.config.retry.max-attempts	6	最大重试次数
spring.cloud.config.retry.max-interval	2000	最大重试间隔
spring.cloud.config.retry.multiplier	1.1	重试乘数
spring.cloud.config.server.bootstrap	false	
spring.cloud.config.server.default-application-name	application	默认服务名
spring.cloud.config.server.default-label		默认标签
spring.cloud.config.server.default-profile	default	默认配置
spring.cloud.config.server.encrypt.enabled	true	
spring.cloud.config.server.git.basedir		Git 仓库地址
spring.cloud.config.server.git.clone-on-start		启动时是否 Clone

(续)

名称	默认	描述
spring.cloud.config.server.git.default-label		默认标签
spring.cloud.config.server.git.environment		环境变量
spring.cloud.config.server.git.force-pull		是否强制拉取
spring.cloud.config.server.git.git-factory		
spring.cloud.config.server.git.password		Git 密码
spring.cloud.config.server.git.repos		Git 中的仓库位置
spring.cloud.config.server.git.search-paths		仓库中的搜索路径
spring.cloud.config.server.git.timeout		超时时间
spring.cloud.config.server.git.uri		Git-URI
spring.cloud.config.server.git.username		Git 用户名
spring.cloud.config.server.health.repositories		
spring.cloud.config.server.native.fail-on-error	false	失败则退出
spring.cloud.config.server.native.search-locations	[]	搜索配置文件的路径。默认与 Spring Boot 应用程序相同,因此 [classpath:/, classpath:/config /, file:./, file:./ config /] 等协议都是可以支持的
spring.cloud.config.server.native.version		
spring.cloud.config.server.overrides		
spring.cloud.config.server.prefix		配置资源路径的前缀
spring.cloud.config.server.strip-document-from-yaml	true	
spring.cloud.config.server.svn.basedir		SVN 的本地仓库目录
spring.cloud.config.server.svn.default-label	trunk	默认标签
spring.cloud.config.server.svn.environment		环境变量
spring.cloud.config.server.svn.password		密码
spring.cloud.config.server.svn.search-paths		搜索路径
spring.cloud.config.server.svn.uri		SVN-URL
spring.cloud.config.server.svn.username		用户名
spring.cloud.config.token		鉴权 Token
spring.cloud.config.uri	http://localhost:8888	配置中心的 URI
spring.cloud.config.username		用户名
spring.cloud.consul.config.acl-token		

(续)

名称	默认	描述
spring.cloud.consul.config.data-key	data	如果格式为 Format.PROPERTIES 或 Format.YAML，则使用以下字段来查找配置
spring.cloud.consul.config.default-context	application	
spring.cloud.consul.config.enabled	true	
spring.cloud.consul.config.fail-fast	true	是否开启快速失败
spring.cloud.consul.config.format		
spring.cloud.consul.config.prefix	config	
spring.cloud.consul.config.profile-separator	,	
spring.cloud.consul.config.watch.delay	1000	
spring.cloud.consul.config.watch.enabled	true	
spring.cloud.consul.config.watch.wait-time	55	
spring.cloud.consul.discovery.acl-token		
spring.cloud.consul.discovery.catalog-services-watch-delay	10	
spring.cloud.consul.discovery.catalog-services-watch-timeout	2	
spring.cloud.consul.discovery.default-query-tag		
spring.cloud.consul.discovery.default-zone-metadata-name	zone	
spring.cloud.consul.discovery.enabled	true	
spring.cloud.consul.discovery.fail-fast	true	是否开启快速失败
spring.cloud.consul.discovery.health-check-interval	10s	执行健康检查的频率（例如 10s）
spring.cloud.consul.discovery.health-check-path	/health	调用健康检查的备用服务器路径
spring.cloud.consul.discovery.health-check-timeout		健康检查超时（例如 10s）
spring.cloud.consul.discovery.health-check-url		自定义健康检查网址覆盖默认值
spring.cloud.consul.discovery.heartbeat.enabled	false	
spring.cloud.consul.discovery.heartbeat.heartbeat-interval		
spring.cloud.consul.discovery.heartbeat.interval-ratio		
spring.cloud.consul.discovery.heartbeat.ttl-unit	s	

(续)

名 称	默 认	描 述
spring.cloud.consul.discovery.heartbeat.ttl-value	30	
spring.cloud.consul.discovery.host-info		
spring.cloud.consul.discovery.hostname		
spring.cloud.consul.discovery.instance-id		
spring.cloud.consul.discovery.instance-zone		
spring.cloud.consul.discovery.ip-address		
spring.cloud.consul.discovery.lifecycle.enabled	true	
spring.cloud.consul.discovery.management-port		
spring.cloud.consul.discovery.management-suffix	management	
spring.cloud.consul.discovery.management-tags		
spring.cloud.consul.discovery.port		
spring.cloud.consul.discovery.prefer-agent-address	false	
spring.cloud.consul.discovery.prefer-ip-address	false	
spring.cloud.consul.discovery.query-passing	false	
spring.cloud.consul.discovery.register	true	
spring.cloud.consul.discovery.register-health-check	true	
spring.cloud.consul.discovery.scheme	http	
spring.cloud.consul.discovery.server-list-query-tags		
spring.cloud.consul.discovery.service-name		
spring.cloud.consul.discovery.tags		
spring.cloud.consul.enabled	true	
spring.cloud.consul.host	localhost	
spring.cloud.consul.port	8500	Consul 代理端口。默认为 '8500'
spring.cloud.consul.retry.initial-interval	1000	初始重试间隔（以 ms 为单位）
spring.cloud.consul.retry.max-attempts	6	最大尝试次数
spring.cloud.consul.retry.max-interval	2000	重试的最大间隔

（续）

名称	默认	描述
spring.cloud.consul.retry.multiplier	1.1	下一个重试时间间隔的乘数
spring.cloud.hypermedia.refresh.fixed-delay	5000	
spring.cloud.hypermedia.refresh.initial-delay	10000	
spring.cloud.inetutils.default-hostname	localhost	默认主机名。用于发生错误的情况
spring.cloud.inetutils.default-ip-address	127.0.0.1	默认 ipaddress。用于发生错误的情况
spring.cloud.inetutils.ignored-interfaces		
spring.cloud.inetutils.preferred-networks		
spring.cloud.inetutils.timeout-seconds	1	超时秒数
spring.cloud.inetutils.use-only-site-local-interfaces	false	仅使用本地地址的接口
spring.cloud.loadbalancer.retry.enabled	false	
spring.cloud.stream.binders		
spring.cloud.stream.bindings		
spring.cloud.stream.consul.binder.event-timeout	5	
spring.cloud.stream.consumer-defaults		
spring.cloud.stream.default-binder		
spring.cloud.stream.dynamic-destinations	[]	
spring.cloud.stream.ignore-unknown-properties	true	
spring.cloud.stream.instance-count	1	
spring.cloud.stream.instance-index	0	
spring.cloud.stream.producer-defaults		
spring.cloud.stream.rabbit.binder.admin-adresses	[]	
spring.cloud.stream.rabbit.binder.compression-level	0	
spring.cloud.stream.rabbit.binder.nodes	[]	
spring.cloud.stream.rabbit.bindings		
spring.cloud.zookeeper.base-sleep-time-ms	50	重试之间等待的初始时间
spring.cloud.zookeeper.block-until-connected-unit		与 ZooKeeper 连接时阻塞的时间单位
spring.cloud.zookeeper.block-until-connected-wait	10	直到连接成功，否则阻塞

(续)

名 称	默 认	描 述
spring.cloud.zookeeper.connect-string	localhost:2181	ZooKeeper 集群连接字符串
spring.cloud.zookeeper.default-health-endpoint		默认健康检查端点
spring.cloud.zookeeper.dependencies		
spring.cloud.zookeeper.dependency-configurations		
spring.cloud.zookeeper.dependency-names		
spring.cloud.zookeeper.discovery.enabled	true	
spring.cloud.zookeeper.discovery.instance-host		预定义的主机可以在 ZooKeeper 中注册自己的服务
spring.cloud.zookeeper.discovery.instance-port		端口注册服务
spring.cloud.zookeeper.discovery.metadata		获取与此实例关联的元数据
spring.cloud.zookeeper.discovery.register	true	在 Zookeeper 中注册为服务
spring.cloud.zookeeper.discovery.root	/services	所有实例都被注册的 ZooKeeper 的根路径
spring.cloud.zookeeper.discovery.uri-spec	{scheme}://{address}:{port}	URI 规范
spring.cloud.zookeeper.enabled	true	启用 ZooKeeper
spring.cloud.zookeeper.max-retries	10	最大重试次数
spring.cloud.zookeeper.max-sleep-ms	500	每次重试最大间隔时间
spring.cloud.zookeeper.prefix		前缀
spring.integration.poller.fixed-delay	1000	
spring.integration.poller.max-messages-per-poll	1	
spring.sleuth.integration.enabled	true	启用 Spring Integration 集成支持
spring.sleuth.integration.patterns	*	默认匹配表达式
spring.sleuth.keys.async.class-name-key	class	
spring.sleuth.keys.async.method-name-key	method	
spring.sleuth.keys.async.prefix		指定前缀
spring.sleuth.keys.async.thread-name-key	thread	
spring.sleuth.keys.http.headers		获取 HTTP 头信息
spring.sleuth.keys.http.host	http.host	
spring.sleuth.keys.http.method	http.method	
spring.sleuth.keys.http.path	http.path	

（续）

名　　称	默　认	描　　述
spring.sleuth.keys.http.prefix	http.	
spring.sleuth.keys.http.request-size	http.request.size	非空 HTTP 请求体的大小（以字节为单位）
spring.sleuth.keys.http.response-size	http.response.size	非空 HTTP 响应体的大小（以字节为单位）
spring.sleuth.keys.http.status-code	http.status_code	
spring.sleuth.keys.http.url	http.url	
spring.sleuth.keys.hystrix.command-group	commandGroup	指定熔断器的命令组名
spring.sleuth.keys.hystrix.command-key	commandKey	指定熔断器的命令名
spring.sleuth.keys.hystrix.prefix		前缀
spring.sleuth.keys.hystrix.thread-pool-key	threadPoolKey	熔断器的线程池名
spring.sleuth.keys.message.headers		
spring.sleuth.keys.message.payload.size	message/payload-size	Payload 类型消息大小
spring.sleuth.keys.message.payload.type	message/payload-type	
spring.sleuth.keys.message.prefix	message/	
spring.sleuth.keys.mvc.controller-class	mvc.controller.class	
spring.sleuth.keys.mvc.controller-method	mvc.controller.method	
spring.sleuth.metric.span.accepted-name	counter.span.accepted	
spring.sleuth.metric.span.dropped-name	counter.span.dropped	
spring.sleuth.sampler.percentage	0.1	采样比例
spring.sleuth.trace-id128	false	生成 128 位跟踪 ID
zuul.add-host-header	false	是否转发请求头
zuul.add-proxy-headers	true	是否转发请求头（代理相关）
zuul.host.max-per-route-connections	20	单个路由可以使用的最大连接数
zuul.host.max-total-connections	200	所有路由可以使用的最大连接数
zuul.ignore-local-service	true	
zuul.ignore-security-headers	true	
zuul.ignored-headers		需要被忽略的 HTTP 请求头
zuul.ignored-patterns		

(续)

名　称	默　认	描　述
zuul.ignored-services		需要被忽略的服务
zuul.prefix		所有路由的公共前缀
zuul.remove-semicolon-content	true	
zuul.retryable		是否支持重试
zuul.ribbon-isolation-strategy		
zuul.routes		路由信息合集
zuul.s-e-c-u-r-i-t-y-h-e-a-d-e-r-s		
zuul.semaphore.max-semaphores	100	Hystrix 的总信号量的最大数量
zuul.sensitive-headers		敏感 HTTP 头不传递到下游
zuul.servlet-path	/zuul	设置 Zuul 路径前缀
zuul.ssl-hostname-validation-enabled	true	
zuul.strip-prefix	true	在转发之前是否从路径中删除前缀
zuul.trace-request-body	true	是否跟踪请求

推荐阅读

架构真经:互联网技术架构的设计原则(原书第2版)

作者:(美)马丁 L. 阿伯特 等 ISBN: 978-7-111-56388-4 定价: 79.00元

《架构即未来》姊妹篇,系统阐释50条支持企业高速增长的有效而且易用的架构原则
唐彬、向江旭、段念、吴华鹏、张瑞海、韩军、程炳皓、张云泉、李大学、霍泰稳 联袂力荐

推荐阅读

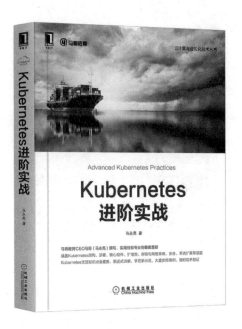

Kubernetes进阶实战

书号：978-7-111-61445-6　作者：马永亮　定价：109.00元

马哥教育CEO马哥（马永亮）撰写，权威性毋庸置疑

涵盖Kubernetes架构、部署、核心组件、扩缩容、存储与网络策略、安全、系统扩展等话题

Kubernetes主流知识点全覆盖、渐进式讲解、手把手示范，大量实操案例，随时动手验证